ABNORMAL PRESSURES
WHILE DRILLING

Origins — Prediction — Detection — Evaluation

elf aquitaine   manuels techniques **2**

ABNORMAL PRESSURES
WHILE DRILLING

Origins — Prediction — Detection — Evaluation

Jean-Paul MOUCHET

M. Sc. NANCY, FRANCE
Ingénieur Géologue ENSPM
ELF UK, LONDON

Alan MITCHELL

B. Sc. NOTTINGHAM, G.B.
Ingénieur Géologue ENSPM
ELF AQUITAINE, BOUSSENS

Boussens, 1989

Cover : designed by Betty Caffier.

MOUCHET, J.P. & MITCHELL, A. (1989). — Abnormal pressures while drilling. — Manuels techniques **elf aquitaine**, 2, 264 p., 145 fig. Elf Aquitaine Edition, Boussens. — ISSN : 0298-7457 — ISBN : 2-901026-28-1

Diffusion : Elf Aquitaine Edition, F-31360 Boussens

© Société Nationale Elf Aquitaine (Production), Boussens
1989

CONTENTS

PREFACE

Abnormal pressure, or at least its superficial effects, has long played an important role in petroleum exploration, and first stirred popular imagination a long time ago.

As early as the 19th century prospectors were fascinated by the abundant oil shows often found in association with mud volcanoes, which are surface manifestations associated with undercompacted layers at high pressure, aligned with anticlinal axes or along fractures. A typical case was described on the Apsheron peninsula in Azerbaijan. Others occur in Romania, the East Indies, Venezuela and Trinidad. There was a tendency to ascribe such "eruptions" to genuine volcanic activity, so some people had no hesitation in linking petroleum with vulcanicity.

It was some 30 years ago that Levorsen, one of the fathers of modern petroleum geology, wrote : "petroleum geology is basically fluid geology". Under his influence, hydrocarbon exploration changed from a static, geometric concept to a more mobilistic and dynamic approach.

Information and observations gathered during the last quarter of a century have only served to reinforce and clarify this "dynamic" view of trapping, whereby oil and gas move naturally towards zones of lower potential and can be stopped by impervious barriers with high internal pressure.

It therefore became important to analyse positive or negative pressure anomalies, and variations in pressure gradients, which Levorsen had described with such foresight as long ago as 1956.

Aquifers thus became a major objective of petroleum specialists, and pressure measurements part of the explorer's toolkit. Some people can well remember how hydrodynamics burst upon the scene in the early 1960's. At the instigation of Hill and Knight in particular, the BRP * laid stress on the notion of hydrodynamic potential in exploration strategy. Ever since that time, traps have been examined not only in their structural aspects but also in their hydrodynamic context.

The study of aquifers gradually led towards understanding the difficult problem of hydrocarbon migration. Little by little it became clear that these various fluid movements were largely governed by basin deformation.

---

* BRP : Bureau de Recherches de Pétrole - a post-war predecessor of Elf Aquitaine.

More recently, sedimentary basins have been studied in the light of plate tectonics theory, and it is now possible to view the petroleum province within an overall geodynamic picture, resulting from an integration of structural deformations, sedimentary changes, heat flow, fluid transfer and so on.

In fact everything is intimately linked and a variation in any one parameter acts on all the others, so constituting the petroleum system.

Positive pressure anomalies play a major role in the origin and concentration of hydrocarbons. Because of their higher potential, such anomalies are able to contribute to the expulsion of fluids, especially hydrocarbons, to facilitate deep fracturing, or alternatively to act as a seal or serve as a lubricant in tectonic deformations. The risks associated with drilling in the presence of abnormally high pressure are also well known, since they include the possibility of a lost well or even loss of life.

This handbook on abnormal pressure makes an important contribution to subsurface techniques. Its value lies in making the practitioner aware of the main tools for detecting and evaluating pressure anomalies during the various stages of exploration, namely before, during and after drilling. The work describes and critically examines methods used. There is no doubt that being aware of these phenomena and acting upon the recommendations which result from them will help to avoid incidents or accidents and contribute to further discoveries.

A. Perrodon

ACKNOWLEDGEMENTS

This book, reflecting as it does the current state of knowledge in fields as diverse as geology, drilling, geophysics or geochemistry, could not have been produced without the help of numerous specialists who gave us the benefit of their constructive criticism during vital but sometimes laborious re-reading and discussion. Among the many people involved we wish to mention those who made active contributions :

A. Chiarelli - J.F. Richy - C. Sejourne	*Fluid Geology*
A. Mallet - P.J. Gallin - D. Grauls	*Subsurface*
A. Lazayres - B. Delbast - A. Vaussard	*Drilling / Mud engineering*
S. Le Douaran - G. Nely	*Basin dynamics*
R. Lanaud - J.M. Drevon	*Geophysics*
J. Connan - J. Espitalie (I.F.P.)	*Organic geochemistry*
F. Walgenwitz - J. Fripiat (C.N.R.S)	*Mineral geochemistry*
Y. Maury - A. Guenot	*Rock Mechanics*
C. Gras	*Wireline logging*
P. Masse - J. Henry	*Tectonics*
A. Ferrand	*Gravimetry*

To all these experts and specialists (and those we have omitted) we address our warmest thanks. A certain number of ideas and illustrations used in this book originate from "in-house" publications. We would like to take this opportunity to list the major sources : J.-L. Rumeau and C. Sourisse, G. Dunoyer de Segonzac, C. Chevalier, P. Valery, A. Chiarelli, D. Grauls, S. Le Douaran, A. Guenot, V. Maury, P. Masse, A. Mallet, P. Rabiller, L. Saugy.

We particularly wish to thank A. Perrodon, who agreed to revise the book as a whole and to write the preface, A. Pierrot who allowed us to devote large amounts of our time to the book, and S. Jardine who gave us his support to the idea of producing this handbook.

Charles Polley translated the text from the original French into English, while D. France revised the translation.

Finally, we will remember the major contribution in the domain of style and layout made by the late Pierre Sauvan.

INTRODUCTION

Technical difficulties are often encountered in petroleum exploration when drilling abnormally pressured zones. Such pressures are a worldwide phenomenon (Fig. 1).

Formation pressure can be abnormally high or low with respect to normal hydrostatic pressure (for definitions and further development see 1.1.1. and 1.1.3.). In this handbook we shall mainly be dealing with abnormally high pressure, generally referred to as overpressure or, more rarely, geopressure.

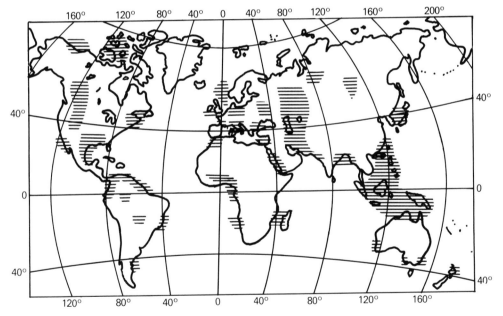

Fig. 1. — World distribution of abnormal pressure.

Most petroleum provinces exhibit abnormal pressure. In fact, abnormal pressure occurs to varying degrees in nearly all sedimentary basins. Whether for technical or climatic reasons, or because significant petroleum yields are unlikely, there is a shortage of data on formation pressure in vast areas of the world (Greenland and Antarctica, for instance). Exploration in areas which are technically more difficult, together with a greater exchange of information, will undoubtedly contribute to a better understanding of formation pressure distribution.

The distribution of observed abnormal pressure is vast, not merely on the geographical scale, but on the vertical scale as well, and can involve the whole sedimentary interval from superficial levels down to depths of 8 000 m and beyond (Oklahoma).

Although more likely to be encountered in recent sedimentary series (see chapter 2), abnormal pressures exist in formations with highly varied lithologies anywhere between the Pleistocene and the Cambrian. There are even examples of abnormal pressure in igneous environments (equilibrium density = 1.8 — Cretaceous andesites, Bai Exi formation — Gulf of Bohai — China).

Recent sedimentary series, which often contain abnormal pressures, are at an active geodynamic stage. Sedimentation dynamics and the consequences of tectonic activity both influence how pressures evolve. Time is a determining factor in this process, conferring a transient, evolutionary character on abnormal pressures. The lifespan of pressure anomalies may vary greatly, from virtually an instant in geological time, anything from a few seconds to several thousand years (landslips, fault movements) to very long timespans involving tens of millions of years (eg. a deltaic zone). An abnormal pressure is not necessarily contemporaneous with the surrounding sediment, as pressure measured in a Palaeozoic formation may for instance have been developed in the Tertiary period.

The presence of a closed or semi-closed environment is an essential prerequisite to the development and maintenance of abnormal pressure. It is the inability of fluid to escape from the pores which conditions the existence and duration of overpressure.

The origins of abnormal pressure differ through the sedimentary, tectonic and physico-chemical processes with which they may be associated. We shall see that some origins lead to abnormal pressures which are detectable before and/or during drilling.

In petroleum exploration, the consequences of abnormal pressures may be both desirable and undesirable. Desirable in the sense that they affect the hydrodynamic gradient and thus encourage hydrocarbon migration. They may also reinforce the efficiency of the seal and thus protect accumulations, or may even have been at the origin of the structure through clay diapirism. Sometimes undesirable, since they are often unpredictable or unquantifiable. Exploration drilling may sustain heavy losses in both human and financial terms because of incomplete knowledge of formation pressure.

In the past there were two recommended methods for drilling wells where there was a risk of abnormal pressure. One was called "drilling for the kick", and consisted in using minimal mud density to achieve a higher drilling rate whilst accepting the risk of encountering a kick. It was then possible to shut in the well, calculate the formation pressure and adjust mud density accordingly. Nevertheless it must be observed that this method often led to uncontrolled blow-outs, and in any event required a high degree of supervision and skill.

The other method, called "overbalanced drilling", consisted, on the contrary, of keeping mud density very high in order to minimize the risk of blow-outs. However, the disadvantages included mud losses, differential sticking, slower drilling rates and suppressed gas shows.

The quality of a drilling programme depends on how well the formation pressure is known. It is not enough to create an inflexible programme and stick to it rigidly. Wherever there is risk of abnormal pressure, the drilling method to be used must consist in continuously evaluating formation pressure as precisely as possible and adapting the drilling programme accordingly. The objective of this handbook is to bring together and explain the differing origins of abnormal pressure, along with methods for predicting, detecting and evaluating such pressure, and the consequent recommendations for wellsite procedures.

This handbook is primarily intended for subsurface geologists and drillers responsible for drilling operations. But it should also be of interest to all petroleum geologists or engineers who need to understand phenomena associated with abnormal pressure. It also fulfils a need to bring together both the knowledge acquired through our experience within the Elf Aquitaine group and that which is to be found scattered throughout the literature. It does not claim to provide the reader with an exhaustive knowledge of the origins of abnormal pressures, but rather to give an insight, to the extent that it is aimed first and foremost at practitioners. The accent is essentially on the various prediction, detection and evaluation techniques available to the reader. The various methods of approach are critically reviewed, and attention is drawn to the necessity of using them in combination, since they may vary in effectiveness from one basin to another.

Related practical themes, such as pressure concepts, gradients, overburden and fracture pressure calculations, are also fully covered.

1. — PRESSURE CONCEPTS

This chapter is devoted to general information on the various types of pressure which the geologist or driller may encounter in the course of forecasting, carrying out or interpreting drilling operations.

Terms are defined and methods of representing pressure are described.

A short section deals with stress concepts, as this is an essential step towards a better understanding of pressure distribution.

1.1. DEFINITIONS

1.1.1. *HYDROSTATIC PRESSURE*

Hydrostatic pressure is the pressure exerted by the weight of a static column of fluid. It is a function of the height of the column and fluid density only. The dimensions and geometry of the fluid column have no effect on hydrostatic pressure.

Fluid column height is considered to be the distance between the measure point and the projection of the well location onto the perpendicular from this point (True Vertical Depth).

It is expressed by the following equation (see section 1.2.1 for units) :

$$P_h = \rho \cdot g \cdot h = 9,81 \cdot \rho \cdot h \tag{1.1}$$

where P_h = hydrostatic pressure (Pascals)
ρ = average fluid density (kg $\cdot$ m^{-3})
g = acceleration due to gravity (m $\cdot$ s^{-2})
h = vertical height of the column of water (metres)

In practice, the following formula is used :

$$P_h = d \times \frac{h}{10} \tag{1.2}$$

where P_h = hydrostatic pressure (kg · cm⁻²) (kg · force)
     d = average fluid density (g · cm⁻³)
     h = vertical height of the fluid column (metres)

The coefficient 10 takes into account metric oilfield units and acceleration due to gravity (9.81).

1.1.2. *OVERBURDEN PRESSURE*

Overburden pressure at a given depth is the pressure exerted by the weight of the overlying sediments. Since this is not a fluid pressure, it is often preferable to distinguish between fluid and matrix by using the term overburden stress.

It may be expressed as follows :

$$S = \rho_b \times \frac{Z}{10} \qquad (1.3)$$

where S = overburden stress (kg · cm⁻²)
     ρ_b = formation average bulk density (g · cm⁻³)
     Z = vertical thickness of the overlying sediments (metres)

The bulk density of a sediment is a function of matrix density, porosity and the density of the fluid contained in the pores.

It is expressed as :

$$\rho_b = \phi \, \rho_f + (1 - \phi) \, \rho_m \qquad (1.4)$$

where ϕ = porosity (from 0 to 1)
     ρ_f = formation fluid density (g · cm⁻³)
     ρ_m = matrix density (g · cm⁻³)

Sediment porosity decrease under the effect of burial (compaction), is proportional to the increase in overburden pressure.

In the case of clays, this reduction is essentially dependent on the weight of the sediments (Fig.2). If clay porosity and depth are represented on arithmetical scales, the relationship between these two parameters is an exponential function. On the other hand, for porosity expressed logarithmically, the porosity-depth relationship is approximately linear.

In sandstones and carbonates, this relationship is a function of many parameters other than compaction, such as diagenetic effects, sorting, original composition and so on.

A decrease in porosity is necessarily accompanied by an increase in bulk density.

In the upper part of the sedimentary column, the bulk density increase gradient is much steeper than at depth (Fig.3). This phenomenon is all the more marked in offshore situations, where the superficial interval consists of water.

Figure 3 illustrates the change in average bulk density with depth.

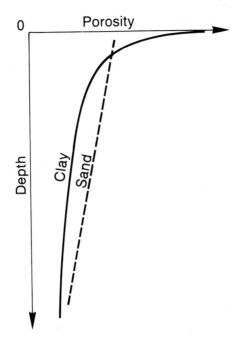

Fig. 2. — Schematic diagram of the porosity/depth relationship.

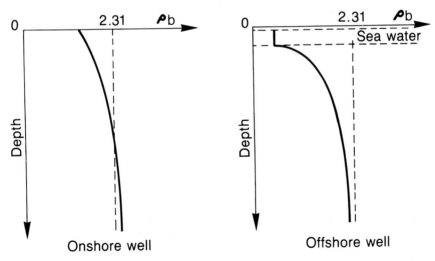

Fig. 3. — Average bulk density changes in sediments (onshore/offshore)
(2.31 = average bulk density at depth).

The average bulk density generally used is 2.31 (equal to 1 psi/ft). This value *can only be used for approximations.* It will be seen later that a more accurate approach requires an evaluation of the overburden pressure interval by interval as a function of differing lithology and density (chapter 4.2).

1.1.3. *FORMATION PRESSURE*

Formation pressure is the pressure of the fluid contained in the pore spaces of sediments or other rocks. It is also called pore pressure (P_p).

Figure 4 illustrates the three categories of formation pressure :

— negative pressure anomaly (or subnormal pressure) : this is pressure which is below hydrostatic pressure : $P_p < P_h$.

— hydrostatic pressure : a function of pore fluid density.

— positive pressure anomaly (or overpressure) : pressure in excess of hydrostatic pressure, and usually limited by the overburden pressure : $P_p > P_h$ (see paragraph 1.3.).

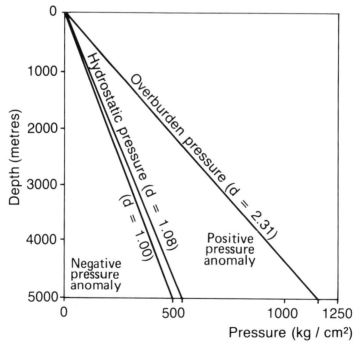

Fig. 4. — Pressure vs. depth plot — classification of pressure categories.

Note that the expressions "subnormal pressure" or "overpressure", which we shall continue to use in line with normal practice, are not strictly correct. It is more accurate to talk of negative or positive pressure anomalies.

1.1.3.1. Normal hydrostatic formation pressure

In normal hydrostatic conditions formation pressure and hydrostatic pressure are equal :

$$P_p = P_h = d \frac{h}{10} \qquad (1.5)$$

In hydrostatic conditions therefore, formation fluid pressure depends on the weight of the column of water saturating the pores of the sediments between the measurement point and the atmosphere. This implies a connection between pores to the atmosphere, regardless of pore morphology and fluid path.

Water density is a function of the concentration of dissolved solids, usually expressed as salinity.

As formation waters vary greatly in salinity, they will also vary in density (Fig.5).

Water type	Salinity Cl⁻ mg/l	Salinity NaCl mg/l	Water density g.cm⁻³
Fresh water	0 to 1 500	0 to 2 500	1.00
Sea water (example)	18 000	30 000	1.02
Formation water (examples)	10 000	16 500	1.01
	36 000	60 000	1.04
	48 000	80 000	1.05
	60 000	100 000	1.07
Salt water saturated in NaCl	192 667	317 900	1.20

Fig. 5. — Water density in relation to salinity (at 20°C, standard conditions).

Generally, surface water densities are of the order of 1 to 1.04 but formation water densities are much more variable and can attain as much as 1.20 and possibly even more in the case of water in contact with evaporites.

The density used in formula (1.2) for evaluating hydrostatic pressure should, where possible, be the average density for the interval between the point being examined and the surface.

The range of average densities generally used for sedimentary basins varies from 1.00 to 1.08.

Examples :

    Nigeria (Niger delta) 1.00
    North Sea (Viking basin)...........1.02
    Gulf Coast (Mississippi delta) 1.07

Average density is determined from pressure measurements (RFT, Tests) and/or from the analysis of produced formation water.

Note that a small variation in water density leads to a significant pressure difference for a given depth :

Example : At 2 000 m — average density 1.00 : P hydrostatic : 200 kg/cm2
               — average density 1.07 : P hydrostatic : 214 kg/cm2

1.1.3.2. Abnormal hydrostatic formation pressure

In certain hydrostatic conditions pressure anomalies can result from variations in h or d.

h has been defined as the vertical height of a column of water (see paragraph 1.1.1). For any given point this height does not necessarily correspond to the vertical height of the well (Z). Similarly, d can be a function of the presence of fluids other than water.

□ *Negative anomaly*

One of the commonest causes is the reservoir outcropping at a lower altitude than the elevation at which it was penetrated during drilling (Fig.6, reservoir A). This explains why such pressure anomalies are so frequently encountered in mountainous areas.

The position of the water table in relation to the land surface (Fig.6, reservoir B) is also a cause of subnormal pressure, especially in arid areas.

It will be noted that both of these pressure anomalies are hydrostatic in origin, but must be taken into account because of their impact on drilling operations.

This is illustrated by figure 6 where :

— the formation pressure in A (upper reservoir) is equal to atmospheric pressure.
— the formation pressure in B (base of the lower reservoir) is a function of water column height h.

It is equal to :

$$P_p = d \frac{h}{10} \qquad (1.6)$$

Another, rarer situation is the marked reduction of average formation fluid density due to the presence of a significantly thick gas column. The shallower the depth of the reservoir in question the more marked will be the effect.

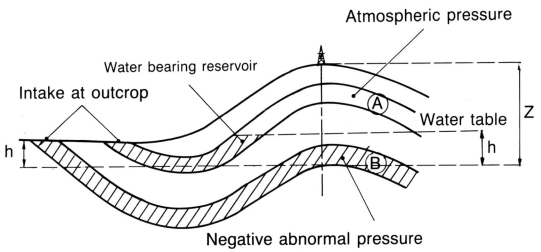

Atmospheric pressure

Water bearing reservoir

Intake at outcrop

A

Water table | Z

h | h

B

Negative abnormal pressure

Fig. 6 - Examples of negative hydrostatic pressure anomalies.

□ *Positive anomaly*

● *Artesian well*

If the intake point (outcrop) of an aquifer drilled is situated at a higher altitude than the wellsite then the formation pressure will be abnormally high (Fig.7).

Since water column height Ze is greater than vertical depth Z of the aquifer in the well, the pressure at A is expressed as :

$$P = \frac{Ze}{10} \times d$$

Intake at outcrop

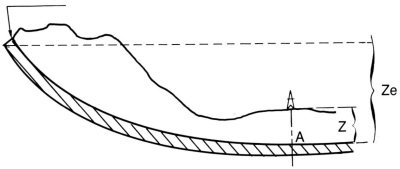

Ze

Z

A

Fig. 7. — Drilling a reservoir below its intake level (Artesian well).

The pressure anomaly due to this difference in height is therefore :

$$P = \frac{Ze - Z}{10} \times d$$

If sufficient information is available on regional hydrogeology it should be possible to prepare adequate mud programmes to counterbalance this type of pressure anomaly.

● *Hydrocarbon column*

Within a hydrocarbon bearing reservoir the fluid column creates a pressure anomaly. This is at its maximum at the top of the reservoir. The force which the water exerts on the hydrocarbon interface due to buoyancy is a function of the differences in density between the water and the hydrocarbons. The resulting pressure anomaly arises from the ensuing gravitational stability. The pressure anomaly at the top of the hydrocarbon column is derived by the following formula :

$$P_{hc} = \frac{h}{10} (d - d_{hc}) \qquad (1.7)$$

where P_{hc} = pressure anomaly at the top of the column (kg · cm^{-2})
　　　h = height of the hydrocarbon column (metres)
　　　d = density of the water (g · cm^{-3})
　　　d_{hc} = density of the hydrocarbons (g · cm^{-3})

Example :

Assume a reservoir 500 m thick encountered at a depth of 2 000 m which is impregnated over a thickness of 400 m by gas of density 0.25 (at bottomhole conditions) with formation water density 1.05 (Fig.8) :

$$P_{hc} = \frac{400}{10}(1.05 - 0.25)$$

$$P_{hc} = 32 \; kg \cdot cm^{-2}$$

For a series at hydrostatic pressure, gas pressure at the top of the reservoir is as follows :

$$P = P_{hc} + \left(\frac{Z}{10} \times d\right)$$

$$P = 32 + \left(\frac{2000}{10} \times 1.05\right)$$

$$P = 242 \; kg \cdot cm^{-2}$$

The chart in figure 9 provides a quick estimate of the pressure anomaly as a function of hydrocarbon column height and differential density between hydrocarbons and formation water.

Overpressure due to this difference in density progressively decreases from a maximum at the top of the reservoir to zero at the water-hydrocarbon contact.

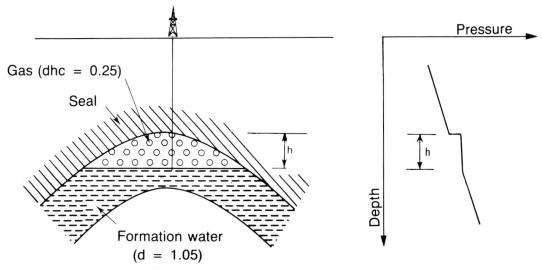

Gas (dhc = 0.25)

Seal

Formation water
(d = 1.05)

Pressure

Depth

h

h

Fig. 8. — Hydrocarbon accumulations — pressure vs. depth plot

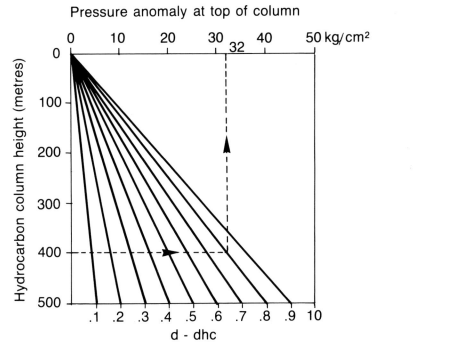

Pressure anomaly at top of column

Hydrocarbon column height (metres)

0   10   20   30 32 40   50 kg/cm²

d - dhc

Fig. 9. — Chart for determining the pressure anomaly at the top of a hydrocarbon column as a function
of its height and the density differential. d_{hc} = hydrocarbon density d = water density

Any abnormal pressure already existing in a series is increased by such an additional pressure anomaly.

Although the hydrostatic pressure anomalies mentioned above are not strictly speaking abnormal pressures, they are generally treated as such because of their implications for drilling operations.

The expression "abnormal pressure" is commonly used in connection with non-hydrostatic (or non-hydrodynamic) positive pressure anomalies. The range of causes of non-hydrostatic pressure anomalies is discussed in chapter 2, "The origins of non-hydrostatic abnormal pressure".

1.2. PRESSURE REPRESENTATION

1.2.1. UNITS OF MEASUREMENT

Even though the official system for units of measurement is the SI International System of Units (SI Units), this system has not been fully adopted by the petroleum industry, in which American companies predominate. In situations where English-speaking influence prevails, US or Imperial measures are often used. Even in countries where SI Units have been adopted officially, the petroleum industry still tends to use more practical units. For example, to simplify calculations the unit of pressure $kg \cdot cm^{-2}$ is still used in preference to the official unit, the pascal. But note that the megapascal ($MPa = 10^6 Pa \neq 10 kg \cdot cm^{-2}$) is destined to become the universal unit for defining stress and pressure in the geotechnical field.

All the conversion tables the reader is likely to need are given in the appendices.

In this book the following units will be generally applied :

— length.. metre ; inch (bit or casing diameters)
— weight .... kilogram force
— pressure... $kg \cdot cm^{-2}$
— density .... $kg \cdot l^{-1}$ (or $g \cdot cm^{-3}$)
— pressure gradient............................. $kg \cdot cm^{-2} \cdot m^{-1} \cdot 10$ (consistent with equivalent mud density)
— flow .... $l \cdot min^{-1}$

1.2.2. PRESSURE/DEPTH REPRESENTATIONS

1.2.2.1. Equilibrium density, equivalent density

The primary aim of drilling mud is to counterbalance formation pressure, which is therefore generally expressed in terms of equilibrium density.

18

The pressure due to a static mud column is said to be hydrostatic ; it is a function of column height and average mud weight (formula 1.1).

Equilibrium density (d_{eql}) represents the average mud weight required to counteract formation pressure.

Equivalent density (d_{eqv}) is the density corresponding to mud column pressure in relation to depth.

— If the mud column is static and its level is at the flowline, equivalent density equals average mud weight (d).

— If the level is below the flowline (well losing mud) :
$$d_{eqv} < d$$

— If the BOP is closed and pressure is applied to the annulus :
$$d_{eqv} > d$$

— While drilling, annular pressure losses and the presence of suspended cuttings in the mud column lead to :
$$d_{eqv} > d$$

In this case equivalent density is called *equivalent circulating density* (ecd).

— While tripping :
 • the swab from pulling the string gives : $d_{eqv} < d$
 • the surge from running in the hole gives : $d_{eqv} > d$

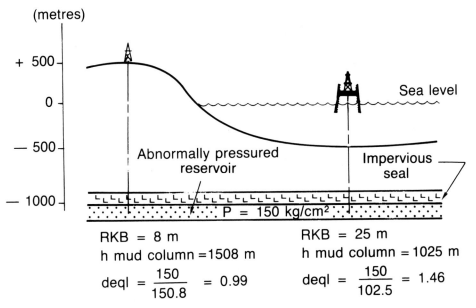

Fig. 10. — Examples of equilibrium density calculations.

RKB = 8 m
h mud column = 1508 m
$$deql = \frac{150}{150.8} = 0.99$$

RKB = 25 m
h mud column = 1025 m
$$deql = \frac{150}{102.5} = 1.46$$

"This means that equivalent density d_{eqv} (not mud weight d) has to be compared with equilibrium density d_{eql} to assess the state of balance of a borehole." (Blowout prevention and control — Chambre syndicale de la recherche et de la production du pétrole et du gaz naturel — Technip Ed. — 1979).

The rotary table (datum level : RKB or drill floor) is always above ground or sea level for drilling. Since the bell nipple is very close to the rotary table, the elevation of the table is used as the datum in calculating equilibrium density.

Figure 10 illustrates equilibrium density calculations for two wells penetrating the same reservoir.

This example clearly shows that the position of the drilling rig plays a significant part in determining the equilibrium density value for the formation, and therefore the equivalent mud weight to be used.

The equilibrium density concept is very practical at the wellsite, but does not completely fulfil the need to obtain pressure vs. depth values which can be compared from well to well and region to region.

1.2.2.2. Pressure gradients

The gradient concept was introduced to give a degree of consistency to pressure data. As we shall see, it must be used with caution.

Formation pressure gradient (G) is the unit increase in pore pressure for a vertical increase in depth of 10 m.

If a pore pressure of 150 kg/cm² is measured at a depth of 1 025 m and a pressure of 165 kg/cm² at 1 125 m, the pressure gradient is given by :

$$G = \frac{165 - 150}{1\ 125 - 1\ 025} \times 10 = 1.50$$

In the case of a formation which is at hydrostatic pressure, within the same reservoir and with phase continuity, the pressure gradient value (known as the hydrostatic gradient) is equal to the *in situ* interstitial fluid density.

Often only one pressure value is known. In such a case the gradient is calculated from the absolute depth (ie. the depth below sea level), so as to make results mutually consistent by referring to the same datum.

If we take the previous example again (Fig. 10) :

Zt (distance between rotary table and sea level) = 25 m

P = 150 kg/cm² at 1 025 m (table elevation)

Absolute depth : 1 025 m − 25 m = 1 000 m

$$G = \frac{150}{1\ 000} \times 10 = 1.50$$

However, in an onshore situation where the top of the water table does not coincide with sea level, calculating pressure gradient relative to sea level will appear to show an anomaly (Fig. 11).

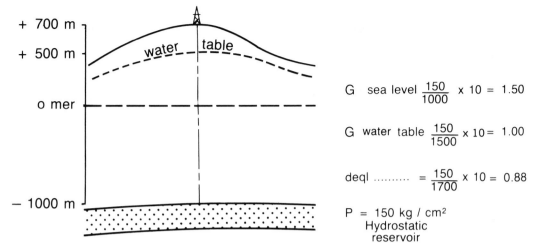

$$G \quad \text{sea level} \quad \frac{150}{1000} \times 10 = 1.50$$

$$G \quad \text{water table} \quad \frac{150}{1500} \times 10 = 1.00$$

$$deql \quad \ldots\ldots\ldots \quad = \frac{150}{1700} \times 10 = 0.88$$

$$P = 150 \text{ kg / cm}^2$$
Hydrostatic
reservoir

Fig. 11. — Apparent gradient anomaly associated with the position of the water table.

In these circumstances the formation pressure gradient should be calculated by reference to the known top of the water table for the region.

While the formation pressure gradient is useful offshore or onshore when sea level and water table are at the same altitude, its use is no longer justified when these two levels differ greatly.

In such cases it is recommended that confusion is best avoided by using only equilibrium density during drilling operations.

Although using gradients usually gives more consistent results and eliminates errors due to topography, it may not be satisfactory if the position of the water table is not accurately known.

Overburden gradient (GG) is the unit increase in stress exerted by the weight of overlying sediments for a vertical increase in depth of 10 m. It is equal to the average bulk density of the sediments throughout the given interval.

Fracture gradient : if the mud density is too high it can cause formation fracturing and consequent mud losses. The upper limit at which a rock forming a borehole wall can withstand pressure from the mud column is called the fracture pressure. The fracture gradient is the unit increase in fracture pressure for a vertical increase in depth of 10 m (chap.4.3).

1.2.2.3. Hydrodynamic levels

Formation fluids can be in a static or dynamic state. Inclined hydrocarbon-water contacts have been discovered, proving that dynamic phenomena play a significant role.

□ *Definitions*

Fluids possess energy which can be expressed as a hydrodynamic potential.

This potential may be represented as a head of water using the following general formula :

$$H = \frac{P \times 10}{d} + Z \tag{1.8}$$

H = hydrodynamic level or head (metres)
P = formation pressure at depth Z $(kg \cdot cm^{-2})$
d = water density
Z = subsea depth of the measure point (absolute depth in metres)

Depending on our knowledge of fluid densities, it is possible to define three types of hydrodynamic level :

• *Pseudo-potentiometric level : d = 1*

Represents formation pressure as a head of fresh water.

This is often applied when the fluid density is unknown — hence the name, "pseudo" potentiometric level.

In the case of an outcropping aquifer it is possible to assume that the pseudo-potentiometric level is given by the altitude of the outcrop,

so that H = Z

• *Piezometric level : d = well measurement*

Represents formation pressure as a head of salt water. The salinity is that measured in a test sample.

The piezometric level is the height at which the water level stabilises in a non-artesian well.

• *Potentiometric level : d = average density*

The density used corresponds to the average density of the water column saturating the reservoir between the intakes and the datum point.

In the case of a fresh water aquifer, the potentiometric level (or true level) corresponds to the pseudo-potentiometric and piezometric levels.

□ *Flow*

Maps of potentiometric levels show that even in deep-lying aquifers hydrodynamic flow occurs. True hydrostatic conditions do not in practice exist at basin scale.

If the potential of a given fluid is not uniform, a force acts upon the fluid to push it in the direction of minimum potential. In fact, fluids migrate away from zones of high potential towards zones of low potential (Fig.12).

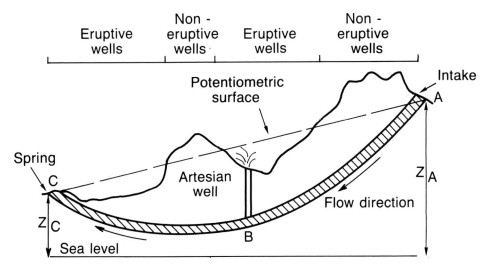

Fig. 12. — Illustration of hydrodynamic fluid flow.

Fluids actually flow from A towards C, even though pressure at B is above zero.

Where the potentiometric surface is tilted there is fluid movement in the reservoir. The direction of tilt reflects the direction of movement (Fig.12).

Since the topographic surface at B is below the potentiometric surface, the well is artesian.

It is possible to define the type of pressure regime by comparing potentiometric and topographic levels.

Differences in the height of hydrodynamic levels reflect the extent to which reservoirs are continuous or limited (Fig.13).

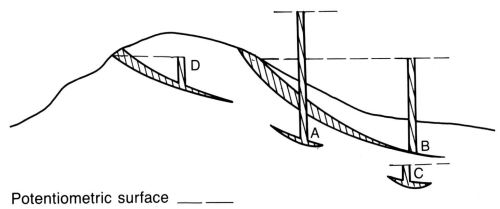

Potentiometric surface _ _ _ _

Fig. 13. — Identifying the pressure regime by reference to the position of the potentiometric surface.

A : Lens with abnormal positive pressure (closed system)
B : Artesian reservoir
C : Lens with abnormal negative pressure (closed system)
D : Reservoir with abnormal hydrostatic negative pressure

Maps of potentiometric levels may also enable us to define the direction of fluid flow.

1.2.2.4. Pressure vs. depth plot

The pressure vs. depth plot is a vital document. It displays not only changes in formation pressure, but also all other data relating to pressure, such as mud weights, overburden pressure, fracture pressure, formation tests, leak-off tests, mud losses and gains (Fig.14).

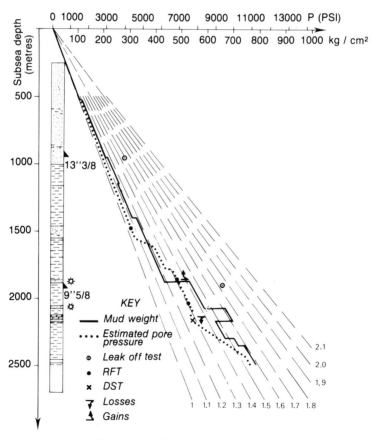

Fig. 14. — Pressure vs. depth plot

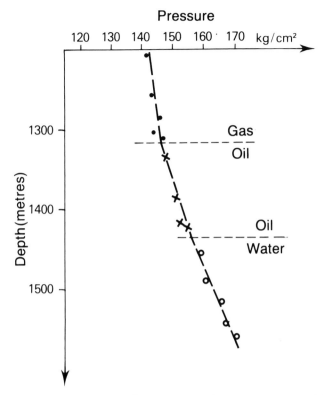

Fig. 15. — Determining water/oil/gas contacts (pressures measured by RFT).

By tracing a group of trend lines onto the pressure vs. depth plot we can read directly either the pressure gradient (with sea level or water table as datum) or equilibrium density (taking rotary table elevation as datum).

One of the most useful applications for the pressure vs. depth plot is evaluating the height of the hydrocarbon column in a reservoir when the hydrocarbon-water contact is difficult to identify by other methods (Fig.15).

1.3. STRESS CONCEPTS

Unlike liquids, which can withstand only internal loads which are equal in all directions (isotropic distribution), solids can support differing loads in a variety of directions.

When a solid is subjected to external forces (for example the forces exerted on a rock sample by the jaws of a press) it reacts by redistributing elementary internal loads, called stresses. These differ in two important ways from the pressures undergone by liquids :

— they differ in spatial direction : a given stress ellipsoid can have any orientation ;
— there are two types. These differ according to how the load is applied. If loading is perpendicular to the elementary surface in question the stress is said to be normal, and can be compressive or tensile (being attributed opposite signs). Tangential loading of the given elementary surface produces what is called shear stress.

Generally speaking, an infinitely small elementary surface within a solid (known as a "facet") is subjected to a small oblique elementary force which can be broken down into a force acting perpendicular to the facet (that is to say a normal stress) and another force parallel to the facet (a shear stress). This is simply a wider application of the classic components of force exerted by a weight on an inclined plane.

A number of items of information are needed in order to define stress conditions at a given point.

The mechanics of continuous environments state that at any point in a solid there exist three planes intersecting at right-angles. Their orientation is unknown, but they are subject to normal stresses only. They are known as the principal planes, and the associated stresses are known as the principal stresses. These planes are therefore not subject to shear stress. This means that six parameters are required in order to describe stress conditions at a point in a solid : the values of the three principal stresses, and the three orientation parameters of the principal planes (Fig. 16). These are much more difficult to determine than the single parameter (pressure) affecting conditions in fluids.

For any given geological structure, there is no reason for stress conditions to obey a vertical and horizontal distribution of the principal stresses, except perhaps in the case of a few "tectonically inactive" basins. Even so, it is to be expected that one of the three principal stresses will be fairly close to vertical, and as often as not it will be assumed fairly accurately that one of them actually is vertical.

In this case borehole ovalisation (due to differential caving) may reveal the azimuth of the two horizontal components.

It must be emphasised that this is only a hypothesis, and that it is very possible to encounter oblique stress fields which are only just beginning to be studied in rock mechanics.

To summarize :

— A hypothetical assumption is made that one of the principal stresses (S_1) is close to the vertical.

— An attempt is made to determine the azimuth of both the horizontal stresses $(S_2$ and $S_3)$ which bear on the vertical facets.

— An attempt is made to determine the values of S_1, S_2 and S_3 (this is difficult to achieve, and it would be better to talk of their order of magnitude). There are very few ways of determining S_1. It is tempting to assume that it is the same as the weight of the overlying

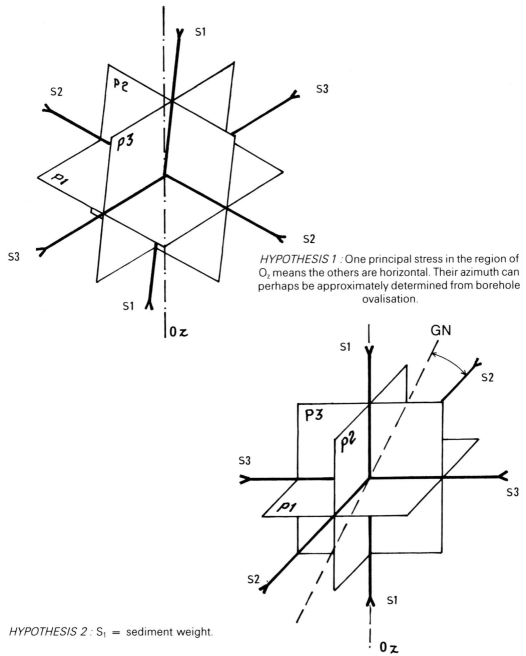

HYPOTHESIS 1 : One principal stress in the region of O_z means the others are horizontal. Their azimuth can perhaps be approximately determined from borehole ovalisation.

HYPOTHESIS 2 : S_1 = sediment weight.

Fig. 16. — Possible stress distribution in a solid.

beds, and it will often be necessary to make this further hypothesis. But in fact there are probably significant differences between the vertical stress component and the weight of overlying beds due to the load-deflecting effects of arches or other structures.

Until quite recently it was believed that both the principal horizontal stresses were around one-third of the vertical stress. This seemed to be borne out by theoretical stress conditions relating to a laterally blocked, heavy isotropic mass. Many observations and measurements have been taken in mines and tunnels showing this to be far from the case. Relative to the vertical stress, it is possible for horizontal stresses to be :

— weaker (known as vertical fields) ;

— similar in magnitude (isotropic fields, wrongly called hydrostatic) ;

— greater or even much greater (4 to 6 times or more : they are then known as lateral, or compressive fields).

At very great depth and for fairly plastic rocks such as salt or marl, stresses tend to equalize and conditions probably become close to isotropic. This hypothesis was proposed by Heim towards the end of last century (1878).

It is false, however, to claim that horizontal stresses at depths shallower than 1 000 m are greater than the vertical stress and that the reverse is true below that depth, because every possibility can be encountered.

This purely intellectual concept of stress has been introduced to give an idea of the internal loads to which a body of rock will be subjected by the forces imposed upon it, and to compare them with the maximum loads it can undergo without excessive fracturing or deformation.

Little is actually known of the forces to which the material in the earth's crust is subjected. Without such forces the deformations which occur would not take place. Such deformation may simply have a lengthwise effect, in which case it is referred to as longitudinal deformation with contraction or extension. If deformation is angular it is known as distortion.

Above all it is important not to confuse stresses (compressive, tensile or shear) with the deformations (contraction, extension or distortion) which they create. The former are expressed in megapascals, the latter as a percentage or "per thousand", indicating how much change of length or form has taken place.

Porous bodies

Until now we have been considering compact bodies. If however they are porous, they can contain a fluid which has a pressure different from the stresses affecting the solid matrix.

The load exerted on a porous solid (S_1, S_2, S_3) is actually distributed over the fluid (if it is prevented from draining away) as well as the matrix. Several theories have been put forward to analyse the respective roles of the forces supported :

— by the matrix : for these forces the concept of effective stress has been advanced. This is a type of average stress affecting the granular structure called σ_1, σ_2 and σ_3,

— by the fluid pressure P_p.

The formulation put forward by TERZAGHI (1923) and confirmed by experiment shows that the effective stresses (σ_1, σ_2 and σ_3) control the deformation of the solid, and that these effective stresses are equal to the total stresses less pore pressure :

$$\sigma_1 = S_1 - P_p \; ; \; \sigma_2 = S_2 - P_p \; ; \; \sigma_3 = S_3 - P_p$$

Other theories, such as that of pore elasticity proposed by BIOT in 1955, which take higher levels of stress into account, propose that pore pressure also plays a role in deformation, and that the above law should be written :

$$\sigma_1 = S_1 - \alpha P_p$$

(for the significance of α see section 4.3).

Generally speaking, pore pressure is limited by the stress conditions in the enclosing formations (the weight of overlying formations and the horizontal stresses, subject to earlier provisos). If pore pressure were to be higher, overlying formations would fracture and pressure would dissipate. It is generally considered that pore pressure cannot be greater than the minimum total stress S_3. Even so there are special conditions whereby the presence of a rigid zone known as a pressure bridge, having exceptional mechanical characteristics such as are found in a dolomite bridge, may make it possible to withstand formation pressure in excess of overburden. Such pressure, sometimes as much as 40 % higher than overburden pressure, is known to occur : eg. 517 kg/cm^2 at 1 600 m in the Checheno-Ingushian Formation, USSR (FERTL, 1976).

2. — THE ORIGINS OF NON-HYDROSTATIC ABNORMAL PRESSURES

Abnormal pressure has many origins. The object of this chapter is to list them and attempt to explain each one in detail. This should give the user enough information to understand the phenomena properly and decide what line of action should be taken when faced with the resulting problems during drilling operations.

Abnormal pressures are hydrodynamic phenomena in which time plays a major role. Every occurrence of abnormal pressure has a limited lifespan, governed on the one hand by the continued existence of the reason for the overpressure, and on the other by the effectiveness of the seal.

A closed or semi-closed environment is in fact essential for abnormal pressure to exist and above all to be maintained. The ability of this environment to resist the expulsion of formation water, implying that drainage is inadequate with respect to time, determines the degree of confinement.

Since it is rare for rocks to be totally impermeable, water trapped in an abnormally pressured environment manages to flow until the overpressure is eventually reabsorbed (HUBBERT & RUBEY, 1959).

Clay, despite very low permeability (10^{-1} to 10^{-7} mD), allows fluid transfer on a geological time scale. Its effectiveness as a seal will depend in particular on its thickness and capillarity. Salt is an example of the ideal seal not only for its plastic behaviour under stress, but also for its perfect impermeability.

Mechanisms leading to abnormal pressure arise in a variety of ways which we shall now examine one at a time.

2.1. THE OVERBURDEN EFFECT (UNDERCOMPACTION)

Before explaining the phenomenon in detail, note that the principle behind the overburden effect arises from the balance between overburden pressure and the ability of a given formation to expel water.

When sediments compact normally, their porosity is reduced at the same time as pore fluid is expelled. Fig.17 shows the established porosity vs. depth relationship for argillaceous

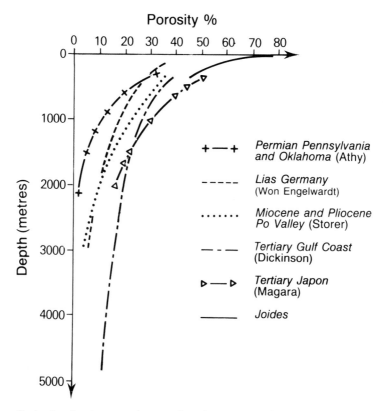

Fig. 17. — Reduction in clay porosity as a function of depth (modified from Magara, 1978).

facies in different regions. The porosity of argillaceous ooze at the sediment-seawater interface can be as high as 80 % (J.O.I.D.E.S. programme). It decreases rapidly with depth within the first thousand metres, reaching average values of 20 % to 30 %. Reduction in porosity is more gradual at greater depths.

During burial, increasing overburden is the prime cause of fluid expulsion. If the sedimentation rate is slow, normal compaction occurs, that is to say that equilibrium between increasing overburden and the ability to expel fluids is maintained.

Laboratory experiments carried out by TERZAGHI & PECK (1948), simulating the role of drainage in clay compaction, made it possible to establish the following simple relationship mentioned earlier :

$$S = P + \sigma$$

where S = overburden pressure (or total stress)
  P = formation fluid pressure
  σ = pressure supported by the matrix (or effective stress)

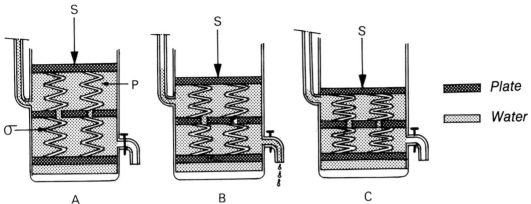

Fig. 18. — Schematic illustration of the compaction model devised by Terzaghi (1948).

Terzaghi's experiment uses a cylinder enclosing three metal plates, the middle one of which is perforated (Fig. 18). The plates are separated by springs, and seal in a certain volume of water. The springs serve to represent the pressure bearing on the matrix. At top left is a conduit fitted with a manometer. At lower right is a drainage tap.

A) Tap closed (no drainage) : when a load S is applied (to simulate overburden pressure) the manometer pressure rises, indicating an increase in fluid pressure P but without causing any deformation. In this case there is abnormal pressure.

B/C) Tap open (drainage) : water escapes, the springs support part of the load and fluid pressure decreases until the springs fully support the load. At C, fluid pressure is hydrostatic (pressure is normal).

The results of the Terzaghi experiments may also be represented as follows :

A : $\Delta S = \Delta P$: the additional load is supported entirely by the fluid
B : $\Delta S = \Delta \sigma + \Delta P$: the additional load is supported by the matrix and the fluid
C : $\Delta S = \Delta \sigma$: the additional load is supported entirely by the matrix.

This model is however simplistic because it does not take into account any reductions in porosity and/or permeability during burial. These happen more gradually than the model suggests. The extent to which application of the load matches the rate of fluid expulsion is the factor governing the magnitude and persistence of fluid pressure in an argillaceous rock.

When the interstitial fluid supports part of the overburden, there is *undercompaction*. This has the effect of simultaneously retarding any reduction in porosity or increase in density.

RUBEY & HUBBERT (1959) put forward an exponential function establishing a relationship between porosity and depth under normal compaction conditions :

$$\phi = \phi_0 \, e^{-cz}$$

where : ϕ = clay porosity at depth Z
        ϕ_0 = surface porosity (Z = 0)
        c   = a constant.

33

The constant c represents the value of the slope of normal compaction (with ϕ plotted on a logarithmic scale).

Later research has enabled compaction vs. depth relationships to be established by reference to wave propagation velocity in clays and shales (see section 3.3.2.2).

Since porosity can vary from 80 % to less than 10 % over a 5 000 m interval, it is easy to see that the volume of water expelled in this way is considerable. It can be demonstrated, using Dickinson's data for the Gulf Coast (1953), that at a depth of 3 000 m the total volume of water expelled is more than 75 % of the original volume of the argillaceous sediment (Fig.19).

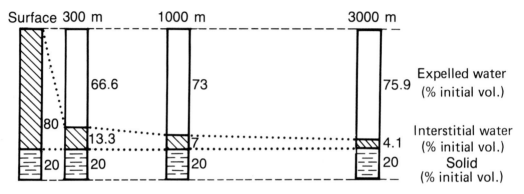

Fig. 19. — Volume of fluid expelled during compaction of an argillaceous sediment.

A reduction in clay porosity is accompanied by an increase in bulk density. Measurements of clay porosity and density form the basis of the study of compaction.

To summarize, normal clay compaction is the result of an overall balance between the following variables :

— clay permeability,

— sedimentation and burial rate,

— drainage efficiency.

The *permeability* of clay is very low, so that water cannot be expelled immediately. For a given sedimentation rate, K_{min} is the minimum permeability that would permit compaction equilibrium to be maintained.

If permeability is below K_{min}, dewatering proceeds more slowly, and abnormal pressure is created which lasts until expulsion is complete.

Similarly, if sedimentation rate exceeds equilibrium conditions, a greater volume of fluid has to be expelled and a higher value of K_{min} is needed to maintain equilibrium. In fact, because permeability does not vary a great deal it is obvious that formation fluid pressure will increase as a consequence. *Pore pressure intensity is dependent on the sedimentation rate.*

Because the sedimentation rate is often greater than that which is needed to allow dewatering of excess fluid, abnormal pressure is very frequent in the following environments : recent deltaic formations, passive continental margins and the accretion prisms of subduction zones, often sites of rapid sedimentation. Generally speaking, the more recent the phase of active subsidence, the greater the probability that pressure anomalies will be encountered.

The probability of abnormal pressure existing increases with the thickness of clay intervals where draining layers of sand or silt are absent.

The *presence of drains* within the argillaceous series is an essential factor governing abnormal pressure. The presence and magnitude of the abnormal pressure appear to be related to the ratio of sand to clay in the sedimentary series. Many authors have discussed the significance of the value of this ratio and its incidence on pore pressure (DICKINSON, 1953 ; HARKINS & BAUGHER, 1969).

Harkins & Baugher (1969) show that when continental sands and clays cover marine clays, abnormal pressure develops *preferentially* in environments with a sand content of less than 15 %.

It will be readily understood that this percentage limit is itself a function of several factors, in particular the degree of confinement of the sand bodies.

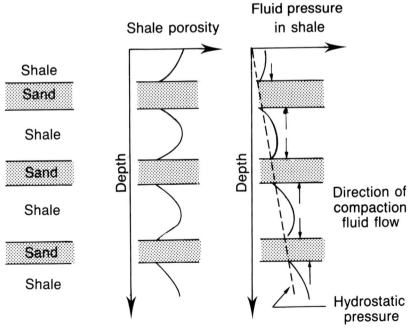

Fig. 20. — Probable porosity and pressure distribution in a shale/sand series during compaction (Magara, 1974, reprinted by permission of the American Association of Petroleum Geologists).

The mechanism for expelling water from clays towards porous reservoirs is the same as that for a fluid to migrate towards zones of lower resistance to flow. As expulsion rate is at a maximum close to drains, the early stages of this process lead to compaction in the immediately adjacent clay beds. The resulting reduction in porosity and permeability retards further fluid expulsion. In certain cases this same mechanism can contribute to the formation of diagenetic cements which affect the sands at the clay boundary.

The fluid pressure within a clay is often assumed to be similar to that in the adjacent sand body with which it is in contact. However, during the compaction process the pressure in the clay further away from the drain is probably higher (Fig.20). This hypothesis, proposed by Magara (1974), seems logical but has never to our knowledge been tested experimentally.

The increases in formation pressure which can be attributed to the effects of sedimentation rate are sometimes insufficient to explain certain pressure anomalies.

Thus we must consider other abnormal pressure generating mechanisms acting separately or in concert.

CONCLUSION

The overburden effect is defined as the result of the action of subsidence on the interstitial fluid pressure of a formation. If fluids can only be expelled with difficulty relative to burial conditions, they must support all or part of the weight of overlying sediments.

Porosity decreases less rapidly than it should with depth, and clays are then said to be undercompacted.

Formation pressure intensity is controlled as much by the rate of subsidence as by the dewatering efficiency. Imbalance between these two factors is the most frequent cause of abnormal pressure.

CLAY DIAPIRISM AND DIFFERENTIAL COMPACTION CLOSELY ASSOCIATED WITH THE OVERBURDEN EFFECT ARE DEALT WITH IN SECTION 2.7.

2.2. AQUATHERMAL EXPANSION

The thermal expansion effect is a concept put forward by BARKER (1972) under the title of "aquathermal pressuring". It is a consequence of the expansion of water due to thermal effects in a closed environment.

If a body of water is raised in temperature its volume increases. If the water is in a hermetically sealed environment, its pressure rises. The extent of the rise in pressure depends not only on the rise in temperature but also on the density of the water.

For example, the pressure rise in water with a density of 1 and a rise in temperature from 50°C to 75°C is 5 600 psi (394 kg . cm^{-2}) (after BARKER, 1972).

Aquathermal expansion only has an effect if the following conditions are satisfied :

— the environment is completely isolated,

— pore volume is constant,

— the rise in temperature takes place after the environment is isolated.

In fact, for the thermal effect to be significant the system must be perfectly closed, since creation of overpressure is associated with a very small increase in the volume of water. The volume increase is in the order of 0.05 % for a burial of 1 km with a temperature gradient of 25°C/km (MAGARA, 1975). This means that even the smallest leak will reduce or even cancel out the thermal effect. Whether the expansion effect gives rise to any overpressure will depend on the extent to which the rate of expansion due to the rise in temperature matches the dewatering rate.

Even so, since the fluid expands so little, clays are usually sufficiently permeable to allow the additional volume to be dissipated in a short geological time given a "normal" geothermal flux. However, if the geothermal gradient steepens significantly and is accompanied by a rapid burial rate, the resulting increase in fluid volume may exceed dewatering efficiency.

Strong thermal anomalies associated with volcanic intrusions or nearby magma chambers may create local overpressure of limited duration (generally less than one million years).

Many objections can be raised against thermal origins of overpressure due to the expansion of water.

— Completely impervious formations are rare.
— Transition zones, which correspond to a gradual shift from hydrostatic to abnormal pressure, reflect true hydraulic transmissivity through clays.
— A rise in temperature reduces viscosity and makes fluid easier to expel.

CONCLUSION

Aquathermal expansion has been proposed as an effect producing increased pressure in sedimentary sequences due to a temperature rise in a closed system.

The effect is governed not only by thermal conditions and water density, but more particularly by the permeability of the environment and the time factor. Its overall contribution is therefore not easy to quantify.

The importance of the thermal effect in the creation of abnormal pressure is a matter for great controversy. Some believe its role is negligible (CHAPMAN, 1980) while others see it as a factor of some significance (MAGARA, 1975a ; GRETENER, 1977 ; SHARP, 1983).

2.3. CLAY DIAGENESIS

Clay transformation and dewatering (with concomitant increase in pore water volume) in the course of diagenesis are often considered contributory factors in the generation of abnormal pressure.

Numerous studies have highlighted the importance of the mineralogical changes which clays undergo during burial.

2.3.1. MINERALOGY

In order to understand these processes more readily, it seems useful at this point to restate some of the pertinent aspects of clay mineralogy.

Argillaceous minerals form part of the phyllosilicates group (the sheet or lattice-layer silicates), which are characterized by alternately arranged sheets of $T_2 O_5$ tetrahedra (where $T = Si$, Al or Fe^{3+}) and octahedra.

The simplest clay mineral, pyrophyllite, is formed by the superposition of two tetrahedral sheets bonded by Al^{3+} ions in the octahedral position.

Its formula is written as follows :

$$Al_2 [Si_4 O_{10}] (OH)_2$$

The structure of pyrophyllite is electrically neutral. The sheets are connected by residual links called Van Der Waal's bonds.

Substitution of Si^{4+} cations in the tetrahedral layer by Al^{3+} creates a negative charge which is compensated by the adsorption of cations and interlayer water. This new structural type is characteristic of, for instance, montmorillonite (smectite family). A strong cation exchange capacity, or water adsorption capacity, gives this type of clay its "swelling" behaviour on contact with water.

The general formula for smectites is written :

$$(Si_{4-x} Al_x)O_{10} (Al_{(2-x)} R_x^{2+}) (OH)_2 EC_x n H_2 O$$

with R^{2+} = Mg, Fe, Mn, Cr etc
    EC = exchangeable cations

Further substitution of Si^{4+} cations by Al^{3+} increases the electrical imbalance and in particular allows potassium or calcium ions to be fixed in an interlayer position. The clay loses its capacity to adsorb water and may gradually change to another type of mineral, illite, which belongs to the mica family. When this transformation takes place the interreticular distance alters from 1.4 to 1 nM (1nM = 10 Å) (Fig.21).

The general formula for illites is written as follows :

$$K_y Al_4 (Si_{8-y} Al_y) O_{20} (OH)_4$$

with $1 < y < 1.5$

Kaolinite is another frequent constituent of clays. It is a purely aluminous variety like pyrophyllite, with the difference that its structure is asymmetric and its interreticular distance is 0.71 nM. It has better thermodynamic stability than the smectites.

The argillaceous phase often encloses complex minerals made up of an alternation of different sheet types, such as illite and smectite, or chlorite and smectite. These are said to be interstratified or mixed layer clays. Relative percentages of the two types of mineral may vary considerably, and they may alternate in a regular or random fashion.

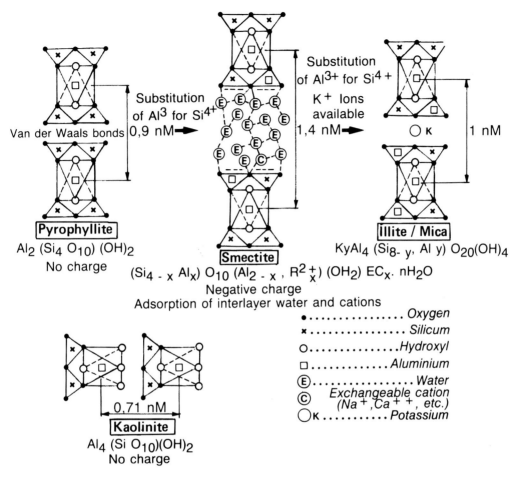

Fig. 21. — The mineralogical structure of clays.

2.3.2. *DIAGENETIC PROCESSES*

Any evaluation of clay diagenesis must be preceded by an analysis of the origin of the clays. "Detrital antecedents must be evaluated and subtracted from the diagenetic effects" (Dunoyer de Segonzac - Elf document).

Research into argillaceous minerals in wells has revealed that as depth increases smectite and mixed layer clays gradually give way to illite (Fig.22).

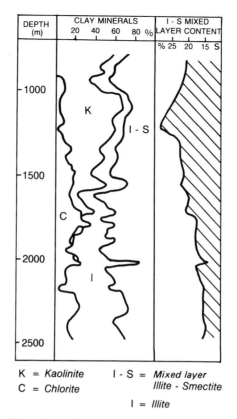

DEPTH (m)	CLAY MINERALS 20 40 60 80 %	I - S MIXED LAYER CONTENT

Fig. 22. — Percentage variation in argillaceous minerals in relation to depth (Ihandiagu - Nigeria).

K = Kaolinite   I - S = Mixed layer Illite - Smectite
C = Chlorite

I = Illite

Unlike the expulsion of excess water during burial and compaction (see section 2.1.), the release of interlayer water from smectites is due to the combined effects of temperature, ionic activity and, to a lesser extent, pressure.

The amount of interlayer water released in this way is dependent on the adsorption capacity of the smectites. This varies according to their composition.

As smectite gradually changes to illite, adsorbed water is expelled in the form of free water. This addition to the pore water can help to generate additional abnormal pressure.

The forces binding water molecules to the silicate layers onto which they are adsorbed diminish as a function of the distance separating them. POWERS (1967) suggests that there are four layers of molecular water in interlayer positions, and up to ten layers on the outside of plates. Several authors (POWERS, BURST, MAGARA) have suggested that the density of adsorbed water can be greater than 1, a phenomenon explained by the very compact structure of the water molecules in the innermost layer.

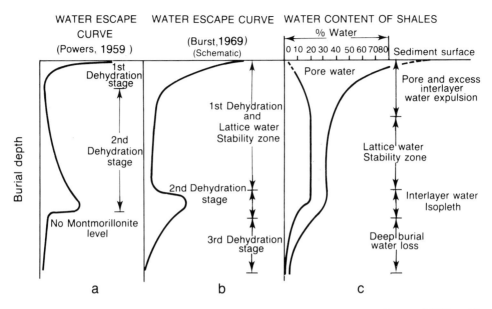

Fig. 23. — Expulsion of water from clays during burial — models suggested by POWERS (1959) and BURST (1969) (in Burst, 1969, reprinted by permission of the American Association of Petroleum Geologists).

POWERS (1959) suggested a two-stage model for the expulsion of water from smectites (Fig.23A) :

1) free pore water expelled near the surface under the influence of pressure ;

2) interlayer water released gradually, first under the effects of pressure, then increasingly under the influence of temperature.

BURST (1969) improved on this model and proposed three stages of dehydration (Fig.23B and 23C) :

1) Expulsion of free pore water and part of the interlayer water, as far as the last two molecular layers, under the influence of pressure. This process takes place increasingly slowly as permeability declines relative to depth.

2) Expulsion of the last-but-one molecular layer of interlayer water under the influence of temperature increase.

3) Gradual expulsion of the last molecular layer.

Burst also showed how the solid and liquid constituents of clay vary quantitatively during the three stages of dewatering (Fig.24).

The depths at which the three stages of the clay dewatering process take place are essentially dependent on the value of the geothermal gradient. Burst and Magara put the temperature at which stage II water release occurs between 90°C and 100°C.

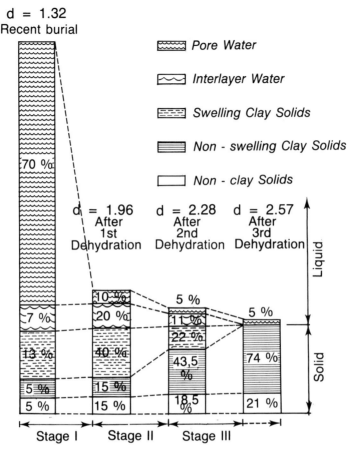

Fig. 24. — Clay evolution during dehydration (Burst, 1969, reprinted by permission of the American Association of Petroleum Geologists).

The chemical composition of the system can also influence the process. A very high concentration of K^+ cations would favour illitization at shallow depths. Conversely any scarcity would retard the process and maintain the interstratified layers to greater depths. The volume of water released at stage II would be 10 % to 15 % of clay volume (BURST, 1969 ; MAGARA, 1975 ; FERTL, 1976). The bulk density of this water is a matter for debate. Powers says 1.4 g . cm^{-3}. This value seems extreme. Burst and Magara prefer 1.15 g . cm^{-3}.

Upon the release of dense molecular water, its volume would increase until overall density matched that of the free pore water. Powers has suggested that this increased volume was the cause of many instances of abnormal pressure in cases where the water could not escape.

There are, however, three areas of uncertainty, namely the quantity of water adsorbed onto the clay sheets, its density, and the temperature range needed for dehydration.

ANDERSON & LOW (1958) and other authors expressed varying views on interlayer water density. However, in the most extreme case the density is thought to be no greater than 2 % more than that of free water.

JONAS et al. (1982), and then FRIPIAT & LETELLIER (1984), who studied the thermodynamic and microdynamic properties of water at or near mineral surfaces, arrived at two conclusions important for current thinking :

— that surface influences affect no more than two or three molecular layers ;

— that the structure of this bound water is not noticeably different from that of free pore water, and it therefore seems improbable that its density could reach the values previously quoted, regardless of its position in the pore spaces (ie. between fine particles or in the interlamellar space).

Regardless of this controversy, it will be noted that the release of water can probably contribute significantly to the creation of abnormal pressure, since it occurs at high temperatures, and therefore at considerable depths where the capacity for water expulsion under the influence of the overburden is reduced. PERRY & HOWER (1972), refer to the transformation of a clay containing :

75 % interstratified layers (made up of 25 % illite and 75 % smectite) into one interstratified layer with 80 % illite and 20 % smectite ; this transformation releases 15.5 % water.

A high geothermal gradient or the confinement of an argillaceous body will both modify clay diagenesis. The abnormally high porosity (and water content) of undercompacted zones explains why their geothermal gradient is abnormally steep. This is a factor which can encourage dewatering and transformation of montmorillonite. On the other hand, abnormal pressure retards dewatering and increases salinity, tending to alter the diagenetic process by comparison with an unsealed environment.

Thanks to the development of techniques for assessing shale cation exchange capacity at the wellsite (see section 3.4.3.3. — shale factor) undercompacted zones have been shown to contain instances not only of montmorillonite increase but also of its decrease.

CONCLUSION

Although clay diagenesis is a contributory factor to abnormal pressure, it is thought to be a secondary rather than a major cause. By adding to the abnormal pressure from overburden effects (undercompaction) it can explain pressure gradients which rise more steeply than the overburden gradient.

2.4. OSMOSIS

Osmosis has been known since the 18th century. It is defined as the spontaneous movement of water through a semi-permeable membrane separating two solutions of different concentration (or one solution and water) until the concentration of each solution becomes equal, or until the development of osmotic pressure prevents further movement from the solution of lower concentration to that of higher concentration (Fig.25).

Osmotic pressure is virtually proportional to the concentration differential. For a given differential it increases with temperature.

The possibility of a clay bed acting as a semi-permeable membrane between two reservoirs with formation water of differing salinity was first suggested in 1933 as a way of explaining the salinity and pressure variations observed in the Gulf Coast.

In 1965 HANSHAW & ZEN suggested that osmosis might contribute to the development of abnormal pressure in a closed environment. Several authors, particularly YOUNG & LOW (1965) and OLSEN (1972), proved experimentally that clay could be considered a semi-

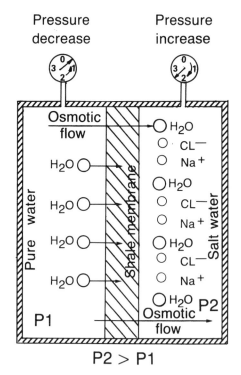

Fig. 25. — Illustration of the osmotic process.

permeable membrane. Its effectiveness in this respect was patchy however, to such an extent that an increased content of very fine quartz in the clay was enough to cause a noticeable reduction in efficiency.

Olsen has demonstrated that the flow of water through a clay bed is dependent on the following four main factors :

— differential pressure,
— differential concentration,
— differential electrical potential,
— temperature.

We should also add to these factors the thickness of the clay, the size of the micropores and the degree of fissuring.

KHARAKA & BERRY (1973) drew attention to the fact that the efficiency of the membrane increases with the cation exchange capacity of the clay.

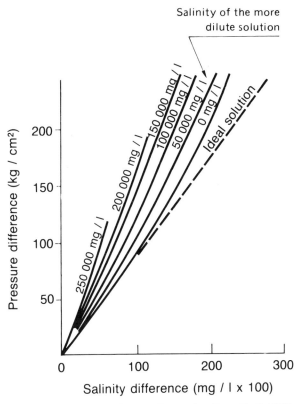

Fig. 26. — Osmotic pressure across a membrane of pure clay (from Fertl, 1976, adapted from Jones, 1968, courtesy of Elsevier Ed.).

In a closed environment, the migration of water towards a reservoir with higher salinity tends to increase pressure in that reservoir until differential pressure is equal to osmotic pressure.

JONES (1968) has drawn an experimental plot to determine differential pressure in terms of differential salinity and the salinity of the more dilute solution (Fig. 26). The plot relates to pure clay. It is intended to give differential pressures as orders of magnitude only, and cannot be applied directly to geological situations. Differential pressure as indicated by Jones can exceed 246 kg/cm^2 (3 500 psi). In theory this is enough to cause formation pressure to exceed overburden pressure.

Osmosis is put forward by several authors to explain certain very rare instances of combined pressure and salinity anomalies, especially the Morrow lenticular sandstones (Oklahoma) where pressure anomalies are sometimes negative and sometimes positive (BREEZE, 1970, in FERTL, 1976).

The process of *reverse osmosis* (or retro-osmosis) consists of the migration of water from strongly saline areas towards areas of weaker salinity under the influence of a pressure differential. This process has been demonstrated in the laboratory. GRAF (1982) gives as a subsurface example the Tertiary formations of the "Chocolate Bayou" field, Texas. This process could explain the anomalies which have been observed there.

It thus seems possible that in certain sedimentary basins fluid flows generated by compaction and gravity may be either accentuated or attenuated by the effects of osmosis or reverse osmosis.

CONCLUSION

Although laboratory tests have proven that osmotic effects are real, the evidence for their existence in nature is far less certain.

It will be noted that laboratory trials used only thin membranes of pure clay and strongly contrasting saline solutions. These cannot easily be extrapolated to the geological environment.

It seems that the capacity for osmosis to generate abnormal pressure is limited to special cases such as sharply contrasting salinity, proximity to salt-domes, and lenticular series. In most instances of abnormal pressure, the role of osmosis is difficult to prove and must be thought of as minor.

2.5. **EVAPORITE DEPOSITS**

Evaporite deposits can have different roles in abnormal pressure :
— a passive role, ie. as a seal
— an active role, ie. as a pressure generator (diagenetic processes).
The role of diapirs is dealt with in section 2.7.

2.5.1. *SEALING ROLE*

Evaporites are totally impermeable, which makes them an almost perfect seal. Because of their inherent plasticity they also have a degree of mobility, and any fractures which occur can repair themselves. This is especially true of rock salt (halite).

During sedimentation, the sealing efficiency of evaporite deposits is a barrier to vertical expulsion of fluid from underlying sediments. If lateral hydraulic conductivity is insufficient for adequate horizontal drainage, the overburden effect will continue to increase and may bring about abnormal pressure in reservoirs and clays alike. This means that argillaceous levels can be undercompacted, as in Northern Germany, where Permian shales underlying Zechstein salt are associated with overpressured gas reservoirs. There are also undercompacted clays in the Saharan Triassic and the Iranian Sudair, as well as in shales intercalated in the Triassic salts of South-West France (Laborde 1 well — Fig.98).

However, on the regional scale this very mobility can jeopardize the effectiveness of the seal. "Mobility is the underlying cause of salt diapirism, causing salt migration and resulting in salt withdrawal structures which, if complete, act like the holes in a sieve. This seems to have happened with the Aptian salts of the Congo" (PERRODON, 1980). Similar situations are observed in the Central Graben and Southern area of the North Sea.

For the distribution of pressures to be properly understood, detailed data needs to be available on the geographical distribution of these "holes in the sieve".

2.5.2. *PRESSURE GENERATION : SULFATE DIAGENESIS*

Gypsum ($CaSO_4$, $2H_2O$) is the only precipitated form of $CaSO_4$ in areas of sedimentation. Transformation into anhydrite or hemihydrate ($CaSO_4$, $1/2 H_2O$) occurs at a very early stage in the burial process.

$$CaSO_4, 2H_2O \text{ (gypsum)} \leftrightarrows CaSO_4 \text{ (anhydrite)} + 2H_2O$$
$$CaSO_4, 2H_2O \text{ (gypsum)} \leftrightarrows CaSO_4, 1/2 H_2O \text{ (hemihydrate)} + 3/2 H_2O$$

Louden (1971) shows that in the presence of pure water and at standard temperatures and pressures, anhydrite is stable above 40°C. Gypsum is stable below that temperature.

KERN & WEISBROD (1964) show that the presence of a salt lowers the threshold temperature for the gypsum-anhydrite transformation, which falls to 25°C instead of 40°C if the water is saturated with NaCl. The role of pressure is also emphasised. As it increases it encourages gypsum dehydration and stabilises the anhydrite-water bond.

When gypsum is transformed to anhydrite, water amounting to 38 % of the original volume is released. Abnormal pressure may therefore develop if this fluid cannot be expelled, but as the process takes place at shallow depth it is usually possible for the excess water to escape.

Similarly, anhydrite rehydration is accompanied by an increase in volume of the same order. Louden suggests that such a process is capable of generating abnormal pressure, as may be the case with the Mississippi Buckner Formation, whose approach marked by a gradual transition from anhydrite to gypsum.

This second hypothesis is still highly debatable. It would require the anhydrite deposits to expand to a considerable extent ; and it has yet to be proven that subsurface anhydrite rehydration occurs on a significant scale.

CONCLUSION

The sealing efficiency of evaporite deposits plays a major role in the generation and maintenance of abnormal pressure. Undercompaction is likely to occur in interlayered or underlying argillaceous series. It is also possible for abnormal pressure to develop in badly drained reservoirs (which are often carbonate in nature) due to their association with evaporites.

Although diagenetic processes cause a significant increase in water volume, the part they play in the creation of abnormal pressure remains to be proven, but is probably only marginal.

2.6. ORGANIC MATTER TRANSFORMATION

The role of hydrocarbons after their migration and trapping was discussed in section 1.1.3.

At shallow depths organic matter contained in the sediments is broken down by bacterial action, generating biogenic methane. In a closed environment the resulting expansion can lead to abnormal pressure. Since it is rare for a good seal to exist at such shallow depths, the gas usually diffuses to the surface. Trapped gas pockets can be a real threat to offshore drilling due to the absence of BOP's in top-hole, but can usually be revealed by high resolution seismic techniques (section 3.3.2.1).

Bacterial activity decreases with increasing depth, gradually giving way to thermochemical cracking. The cracking process involves transforming a heavy product into a lighter one under the influence of high temperatures. The JOIDES project revealed ethane at a depth of 250 m, proving that thermal cracking begins at an early stage (Claypool *et al.*, 1973), as ethane is not produced in significant quantities by bacterial action.

Thermochemical generation of light hydrocarbons such as methane proceeds at an increasing rate as temperature rises. It reaches a maximum above 100 °C to 120° C, and continues until carbonised kerogens are produced (Fig.27).

The cracking process creates hydrocarbons from organic matter and also produces light hydrocarbons from heavy ones. The transformation increases the total number of molecules, and therefore the volume they occupy. If this takes place in an "open" environment, no increase in pressure occurs. If on the other hand it occurs in a "closed" or "semi-closed" environment, it can cause pressure to rise. This will depend on the degree to which the environment is confined and the final nature of the hydrocarbon product.

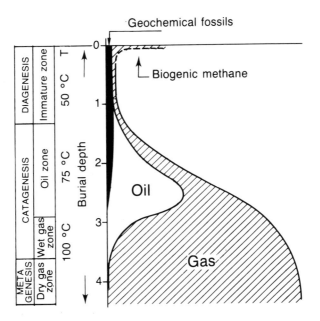

Fig. 27. — Hydrocarbon generation as a function of temperature and depth (modified from Tissot & Welte, 1978, and Kartsev, 1971).

As compaction proceeds and less water is expelled, decomposing organic matter would tend to cause the water to become saturated in gas and eventually produce free gas. If this gas is unable to escape it causes abnormal pressure (HEDBERG, 1974). Pressure anomalies and undercompaction due simply to the overburden effect will be magnified if gaseous hydrocarbons are generated at the same time. Many authors agree that the rise in pressure may lead to microfissuring and allow pressure to be partially dissipated, thus contributing to primary migration.

Undercompacted clay zones often have a high gas content. This suggests that cracking of the organic matter makes a contribution to abnormal overburden pressure. On the other hand, since some undercompacted clays have no sign of gas, it can be assumed that hydrocarbon transformation is not the dominant cause of abnormal pressure.

CONCLUSION

Over and above the fact that undercompaction is often accompanied by gas shows rich in heavy components, thermal cracking of organic matter can be a cause of abnormal pressure in its own right. It can develop in either shaly sand series or carbonate series, provided organic matter is present and the system is sufficiently confined.

2.7. TECTONICS

In general, where deformations occur due to tectonic stress, they cause modifications in fluid pressures and in the distribution of masses. This means that tectonics may create positive pressure anomalies or restore pressure to normal.

The link between tectonics and fluids can be viewed from two related standpoints :

— tectonic activity causes rock deformations which have a direct or indirect effect on fluid pressure distribution ;

— to a greater or lesser extent fluid pressure alters the way in which deformations develop as a result of stress.

In any event, tectonic activity can have a variety of effects. We therefore need to distinguish between a number of different cases, at the same time realising that they may occur together.

2.7.1. *RELIEF AND STRUCTURING*

Changes in formation relief and geometry are a direct cause of pressure redistribution. Relief induces hydrodynamic activity, which in turn is an underlying cause of some of the pressure anomalies observed (see section 1.1.3.).

Some authors also suggest that a deep-lying series may be uplifted and part of the overlying strata then eroded. In this way zones of high pressure could be brought closer to the surface, which would make them appear anomalous. Such situations are referred to as *palaeopressures.*

This hypothesis assumes a closed system and rapid uplift, but raises numerous objections. Since tectonic movement such as this would usually be accompanied by fracturing, pressure would tend to dissipate. In addition, the lower temperature at the reduced depth would decrease the fluid volume, and therefore the pressure.

In fact temperature equalisation probably ensures that fluid pressure declines more quickly than overburden pressure during erosion, thus leading to a negative pressure anomaly (MAGARA, 1975).

2.7.2. *REORGANISATION OF STRESS FIELDS*

The immediate effect of tectonic activity is to modify the force and direction of the stress field. Sediments are subjected not only to the overburden stress of their own weight, but also to tectonic stress.

If fluids are able to escape, compaction is likely to be faster than under the influence of burial only.

Note that in a tectonic stress field, vertical forces are no longer the main influence on compaction. This can be shown in calcareous series subjected to compressive tectonic stress, where stylolites possessing oblique or even horizontal peaks are seen to occur. On the other hand, if such series are compacted by overburden effects or subjected to tectonic extension, stylolites exhibit vertical peaks.

Tectonic action can set up stresses so rapidly that fluid expulsion is hampered, and this can generate overpressure. Hydraulic fracturing may, however, ensue, dispersing wholly or partially any pressure anomalies thus created.

It should be noted that if the cementing of fissures and pore spaces occurs at the same time as a reduction in porosity due to tectonic compaction, permeability can become unevenly distributed and encourage formation of localised pressure anomalies.

Overthrust zones

In some cases abnormal pressure is clearly associated with overthrust faulting (eg. foot-hills, as in Canada, the Apennines, the Colombian Llanos, etc.). Fluids at high pressure and temperature act as a lubricant for the movement of the overthrust block (RUBEY & HUBBERT, 1959 ; GOGUEL, 1969). Very pronounced overpressure can be induced by contact between the overthrust surface and the substratum. Rapid loading occurs, causing abnormal pressure in underlying confined sequences. The significance of these effects will depend on the thickness of the nappe and the degree of hydrodynamic confinement within the sequences beneath the overthrust.

2.7.3. *FAULTS AND FRACTURES*

The effect which faults have on fluid pressure distribution depends on many factors (Fig.28) :
— whether they form an effective seal or on the contrary act as a drain,
— how they displace reservoirs and sealing strata,
— the original distribution of sealing and reservoir sequences.

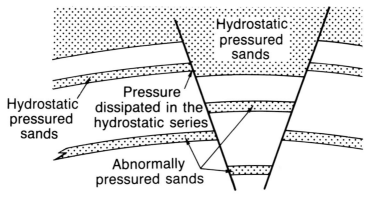

Fig. 28. — The role of faults in pressure distribution.

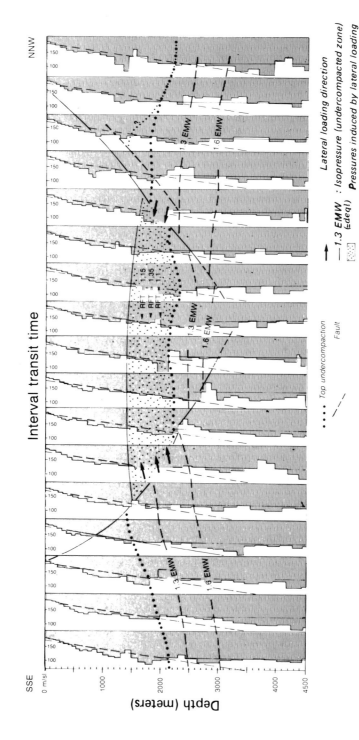

Fig. 29 — Example of laterally induced overpressure - Growth fault zone, Niger Delta.

Normal faults are the result of a stress field where S_1 is vertical and S_3 is horizontal. As they are created by a system in extension and therefore tend to be open, they are often effective drains, and provide links between reservoirs which help to equalise pressure gradients. However, in the presence of saturated fluids the fault plane becomes, due to the localised pressure decrease, a site for syntectonic or premature crystallization of calcite, quartz, anhydrite or dolomite, none of which is very permeable. If this happens, faults will act as a barrier or seal to a reservoir.

Reverse faults are the result of a stress field where S_1 is close to horizontal and S_3 nearly vertical and are thus more likely to be closed. In very broad terms they tend to be a barrier to fluid circulation, either in their own right or because of the alterations they engender in surrounding formations.

Tear faults are the result of a stress field where S_1 and S_3 are horizontal and S_2 is vertical. As with normal faults, whether they act as barrier or drain depends on whether there is syntectonic mineral crystallization. Their impact will also be affected by the relative displacement of the compartments on either side of the fault.

Fault displacement is also an essential factor in the distribution of fluid pressure. If a fault is to isolate a section of reservoir, it needs to displace its walls in such a way as to bring the porous layer into contact with an impervious zone. If the movement brings reservoirs into contact at some point, pressure conditions in the two compartments will equalise.

Major faults, especially strike slip faults, create fracture corridors or zones which act as a drain as long as the fractures are not sealed by mineralisation.

Joints are *fractures* with little or no displacement. They play a dominant role in pressure systems. They are capable of depriving impervious rocks of their ability to act as a seal. On the other hand, plastic clays, anhydrites and above all salt deposits are self-repairing, and are the only seals capable of retaining their impermeability even in conditions of severe deformation (Iraq, Iran). Fracture intensity depends on both the stress field (ie. the type of tectonic activity) and the mechanical behaviour of the layers.

It is important to note that overpressure may often be laterally induced by the juxtaposition, by fault movement, of formations with different pressure regimes. A typical case is figure 29, which illustrates the transmission of pressure via a growth fault in the Niger Delta. The seismic velocities in this case indicate the top of undercompaction to be much deeper than the first abnormal pressures encountered.

2.7.4. *TECTONICS AND SEDIMENTATION*

2.7.4.1. Deltaic areas

The development of a delta depends on the balance between sedimentation rate, subsidence rate and eustatic variations in the sea level. Such environments encourage the formation of undercompacted zones in locations which are determined by the sedimentology of the area. They form in underdrained or undrained parts of the delta.

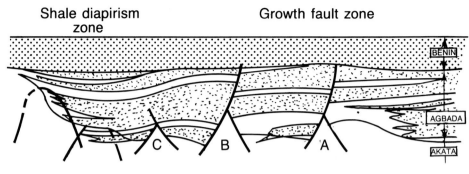

S   N

Shale diapirism zone   Growth fault zone

A, B, C, : *Shale ridges*

Fig. 30. — N-S schematic cross-section of the Niger Delta (Rio del Rey — Cameroon).

In such deltas as the Niger and the Mississippi it is possible to observe the following features, which depend on the direction of sediment flow (Fig.30) :

— a proximal zone, where growth faults will develop preferentially,

— a distal zone with shale domes and ridges.

☐ *Growth faults*

Growth faults, also known as synsedimentary or listric faults, possess a curved fault plane which is invariably concave towards the basin. This plane is nearly vertical in its upper part, then tends gradually to conform to the dip of the strata as its slope decreases towards its base (Fig.31). The downstream compartment displays thickening of the sediments in the form of a "roll-over" (compensation anticline) near the fault.

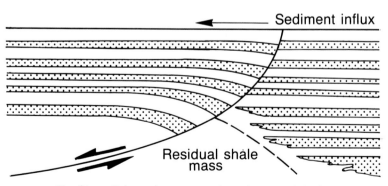

Fig. 31. — Schematic cross-section of a growth fault.

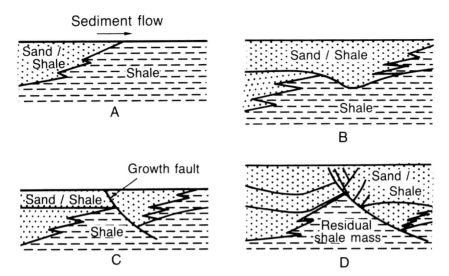

Fig. 32. — Diagrammatic development of a residual shale mass
(Bruce, 1973, reprinted by permission of the American Association of Petroleum Geologists).
A, B, C, D : successive stages of the mechanism.

Although the importance of gravity in the development of growth faults is undisputed, trigger mechanisms are still open to debate. Basement tectonics, gravitational slumping of the sediments, salt or clay diapirism, differential compaction or a combination of these factors have all been suggested. CRANS et al. (1980) showed that during compaction, clays could slide under their own weight on a slope of less than 3°. Lowering of the downdip compartment creates a surface depression which traps sediments. Their additional weight encourages further slipping. The slip plane is itself seated in an incompetent layer.

The base of the updip compartment of growth faults often includes a ridge of undercompacted shale (residual shale mass) resulting from differential compaction (Fig.32) (BRUCE, 1973 ; EVAMY et al., 1978).

This association of growth faults and shale ridges has been revealed in the Niger delta, where it occurs on a regional scale (Fig.30).

The preferential site for hydrocarbon accumulation is the roll-over structure of the downdip compartment against the fault. If such structures are drilled, there is always the risk of crossing the fault and penetrating the ridge of undercompacted shale, thus risking a sudden rise in formation pressure.

□ *Shale diapirism*

Shale domes are the result of intrusive flow from underlying layers (shale diapirism). They are always undercompacted, and therefore abnormally pressured. Figure 33 shows an example of clay diapirism.

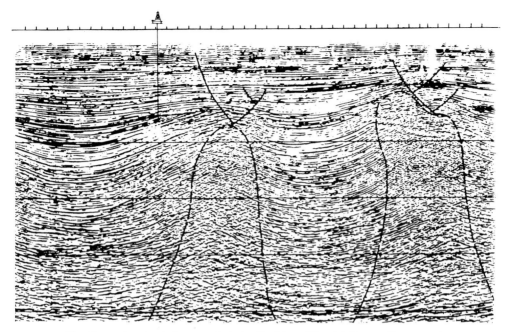

Fig. 33. — A seismic image of shale diapirism (Rio del Rey — Cameroon).

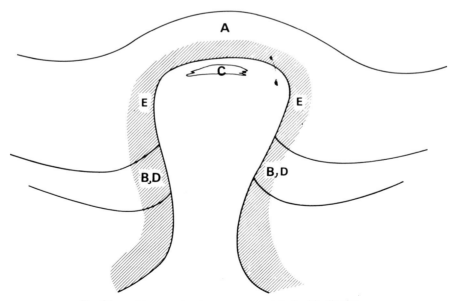

Fig. 34. — Abnormal pressures associated with diapirs.

Shale domes are formed by processes similar to those which form salt domes, and the following pressure anomalies are likely to be generated (Fig.34) :

- palaeopressure due to uplifting previously deep-lying formations to shallower depths (A),
- confinement of pierced layers (B),
- isolated "rafts" on the top of the diapir (C). Because the overburden pressure transmitted to such isolated formations is omnidirectional, significant overpressure will develop within them (salt domes),
- pressure transfer from the undercompacted clays to the pierced reservoirs (D),
- osmotic effects due to raised salinity in the water of formations close to the salt dome (E).

2.7.4.2. Subduction zones

Argillaceous sediments are often buried rapidly in geosynclinal zones and in subduction zones where two tectonic plates converge. An example is the subduction of the Atlantic plate beneath the Caribbean plate. In this particular case, very fine grained sediments from the Orinoco and Amazon deltas accumulate on very thick beds in the rapidly subsiding Caribbean arc foredeep. They are rapidly buried and come under the compressive deformation of the tectonic accretionary prism (Fig.35).

Undercompacted argillaceous layers are favourable to the development of overlying deformation because they act as lubricants, amplifying the movement (MASCLE et al., 1979). Décollement, the final compressive stage allowing overthrust to occur, depends on frictional forces at the base. These forces are cancelled out in an incompetent argillaceous environment, as undercompaction facilitates overthrusting (CHAPMAN, 1974).

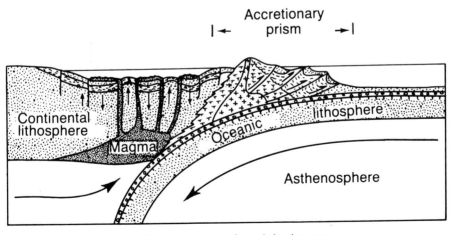

Fig. 35. — Example of a subduction zone.

The Barbados accretionary prism is an example of multiple imbricated overthrust faults. Their development is intimately linked with abnormal fluid pressures in undercompacted argillaceous layers (SPEED, 1983). Some décollement planes are visible on seismic sections (WESTBROOK et al., 1982). Extensive clay diapirism with associated mud volcanoes confirms this undercompaction (WESTBROOK & SMITH, 1983 ; VALERY et al., 1985).

Mud volcanoes are the ultimate manifestation of clay diapirism. They tend to be situated along large, active transcurrent faults, such as in New Zealand (Alpine fault), Trinidad, the Caspian Sea, Azerbaijan, and so on. If gas is present it intensifies the process by increasing differential pressure (HEDBERG, 1974).

CONCLUSION

Tectonics and fluid pressures interact to give a variety of effects. The list given in the text is by no means exhaustive.

Nevertheless, it is possible to summarize tectonic mechanisms in outline as follows :
Extension → open fractures → pressure dissipation

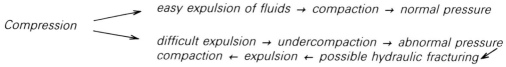

easy expulsion of fluids → compaction → normal pressure
Compression
difficult expulsion → undercompaction → abnormal pressure
compaction ← expulsion ← possible hydraulic fracturing

2.8. MISCELLANEOUS

2.8.1. *CARBONATE COMPACTION*

By virtue of their texture, carbonates do not generally undergo the effects of undercompaction seen in clays and shales. Chalk is an exception, being made up of coccoliths which tend to take up a horizontal arrangement during compaction. This special texture makes chalk behave rather like clay with respect to porosity reduction and water expulsion during burial. These pelagic carbonates are deposited slowly, and their initial porosity is around 70 %. This porosity is gradually reduced to a value of between 5 and 10 % at 3 000 m. Very thick chalk deposits may develop undercompaction because of their low permeability.

When porosity declines to a level of 35 % or less, mechanical compaction is replaced by "chemical compaction" (ie. processes involving pressure-solution). At this stage the coccoliths dissolve at their points of contact and $CaCO_3$ is precipitated in the pore spaces, with the result that porosity and permeability are diminished (SCHOLLE, 1978).

Figure 36 shows the unusually high porosities observed in the abnormally pressured chalks of the North Sea and Gulf Coast.

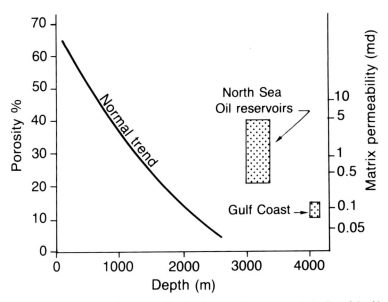

Fig. 36. — Porosity and permeability anomalies in abnormally pressured chalks of the North Sea and the Gulf Coast (after Scholle, 1978).

2.8.2. *PERMAFROST*

When water changes into ice its volume increases. Water contained in surface sediments of permafrost regions is frozen, but in certain conditions, pockets of ground surrounded by permafrost can exist in an unfrozen state. Such pockets are known as taliks.

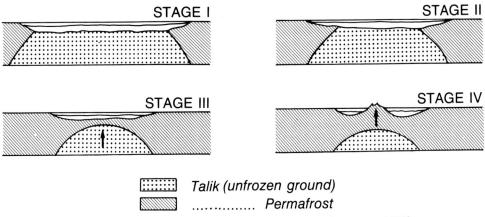

Fig. 37. — Diagram showing how pingos develop (Gretener, 1969).

But ice is quite impermeable, so that if a talik does freeze, permafrost impedes expansion and encourages abnormal pressure to develop.

An example quoted by GRETENER (1969) refers to the pingos of the Canadian Arctic.

The development of pingos can be described schematically as follows (Fig.37) :

1) if a lake in a permafrost environment is deep enough to ensure that the water at the bottom is not frozen, a talik zone can exist beneath it ;

2) as the lake gradually silts up in summer its depth is reduced until it is able to freeze to its full depth in winter, and the talik can then be invaded by permafrost ;

3) eventually a permafrost bridge develops, isolating the talik ;

4) if the talik freezes, its water increases in volume. This leads to an increase in pressure which uplifts the permafrost bridge, forming pingos.

Although the phenomenon is very localised, it must be taken into account when drilling in regions of permafrost.

2.9. CONCLUSION

This account of the various ways in which abnormal pressure can arise has attempted to distinguish between major and secondary causes. Identifying the cause is generally a delicate matter, and calls for a sound knowledge of the geology of the region. The crucial importance of seals and drains in developing and maintaining abnormal pressure has been demonstrated. Time is the determining factor in fluid dispersal, which explains why abnormal pressure is more commonly found in association with young sediments. High pressure may result from a combination of various causes, and these are more likely to be found in clay-sandstone sequences because of the mechanical, physical and chemical properties of clays. The lithological changes which some of the causes bring about can be used for detection purposes during drilling operations. The characteristics and typical environments of the various origins are summarised in the following table.

THE ORIGINS OF ABNORMAL PRESSURE : SUMMARY TABLE

The table below summarizes the various origins and characteristics of abnormal pressure.

Origin	Characteristics	Environment
Overburden effect	— major contribution to the existence of abnormal pressure — leads to undercompaction — geographically widespread — long-lasting effect linked to sedimentation rate	Young clay-sand sequences : • deltas • passive continental margins • accretion prisms of subduction trenches • evaporite deposits
Aquathermal expansion of water	— requires a very well sealed environment — temperature plays a major role — may be superimposed on the overburden effect	Closed system with steep geothermal gradient Volcanic zones
Tectonics	— very varied characteristics due to redistribution of masses and fluid pressures	Faults, folds, overthrust faulting, clay diapirism, salt diapirism Lateral pressuring
Cracking of organic matter and hydrocarbons	— cracking = increased volume — develops either in undercompacted environments or independently — important role of temperature	Sediments rich in organic matter
Clay diagenesis	— second order cause. May be superimposed on the overburden effect — geothermal gradient plays a major role — significant smectite proportion in the original deposit	Thick argillaceous sequences
Osmosis	— rare second order cause — transient, unstable phenomenon — difficult to prove	Interlayering of clay with lenticular reservoirs of contrasting salinity
Miscellaneous : Sulfate diagenesis Carbonate compaction Permafrost	— special cases : localised, transient phenomena	Evaporite deposits chalk talik/permafrost

3. — PREDICTION AND DETECTION

There is always an element of risk with any drilling operation, but the presence of abnormally pressured sequences can significantly increase these difficulties. In such conditions, successful drilling requires the use of every means of detection at our disposal.

These means are not universally applicable, and their effectiveness varies from case to case.

A number of methods are available for the qualitative or quantitative assessment of abnormal pressure. The first concern must be to study local structure and lithostratigraphy to reveal any closed system which may be present. This initial phase may be able to detect zones of potential risk, and *must be incorporated into the preparatory stages of the drilling programme,* even though it can give no guarantees about the presence and magnitude of abnormal pressure.

Given all the predictive methods available, successful drilling still depends on the effectiveness of the methods adopted at the wellsite and on the way they are used in combination.

The regional studies carried out in advance of locating wells will not be discussed in detail here. On the other hand we must emphasise the importance of this investigative work, and the absolute necessity for the subsurface geologist to be aware of it, as it will help in the interpretation of wellsite observations.

Before reviewing the various methods for detecting abnormal pressure, we shall discuss what is meant by a normal compaction trend and describe the characteristics of undercompacted zones.

3.1. NORMAL COMPACTION TREND

In order to evaluate abnormal pressures linked to compaction anomalies it is necessary to define a normal compaction trend for reference purposes.

Compaction represents a reduction in porosity with increasing depth, and as we saw in section 1.1.2., a linear relationship is revealed on a logarithmic plot of porosity vs. depth. Argillaceous facies must be used for determining this relationship. Pure clays with identical mineralogical composition and texture theoretically show a unique compaction trend.

In reality clays and shales have widely varying facies. The slope of a normal compaction trend curve is influenced by the following items in particular :

— the mineralogy and relative proportions of the phyllosilicates in the clay,
— the non-argillaceous mineral content (quartz, carbonates, organic matter etc),
— the sedimentation rate, which conditions the texture by means of the spatial arrangement of particles. Porosity is lower if sedimentation occurred at a slower rate (MORELOCK, 1967 ; PERRY, 1970),
— the geothermal gradient (REYNOLDS, 1973).

The variety of porosity vs. depth relationships is explained by the incidence of these different causes (Fig.17). In practice it is advisable to establish normal compaction trends on at least a regional basis.

3.2. CHARACTERISTICS OF UNDERCOMPACTED ZONES

3.2.1. *TRANSITION ZONE*

We have already seen the importance of fluid movement during compaction of a clay-sand sequence (section 2.1). In accordance with the laws of hydrodynamics such movements are away from zones of high potential towards areas of lower potential. In the

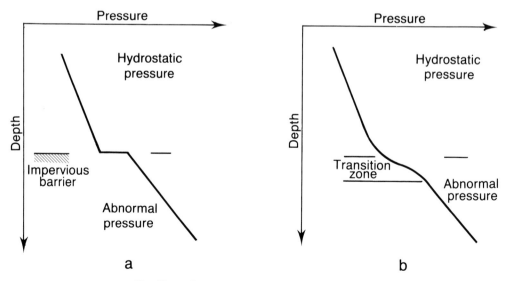

Fig. 38. — Pressure changes — Transition zone.

case of abnormally pressured zones, fluids tend to move towards formations under hydrostatic pressure.

If there is a seal between two zones of different potential, it prevents any communication between them and shows up as an abrupt change in pressure conditions (Fig.38a).

On the other hand, if the seal is not perfect, escaping fluids will bring about a partial dispersal of pressure in the upper part of the abnormally pressured sequence. Such an interval exhibits a gradual change in formation fluid pressure from abnormal to hydrostatic, and is usually called a *transition zone* (Fig.38b).

Fluid pressure changes in clay deposits correspond to an altered state of compaction resulting in porosity and density variations (Fig.39).

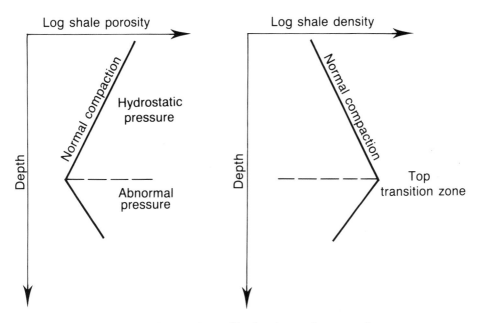

Fig. 39. — Schematic profile showing undercompaction.

The thickness of the transition zone depends on clay permeability, drainage conditions and the time factor. In the case of monotonous argillaceous sequences, this can be a considerable thickness (several hundred metres is not uncommon).

It is easier to detect abnormal pressure if the transition between the different pressure zones is gradual. This means that the thicker the transition zone, the easier the undercompacted zones will be to detect.

3.2.2. *DIAGENETIC CAP-ROCK*

Various authors refer to the presence of indurated and carbonated shale levels at the top of certain undercompacted zones. In many text books such levels are called chemical cap-rocks or diagenetic cap-rocks (Fig.40).

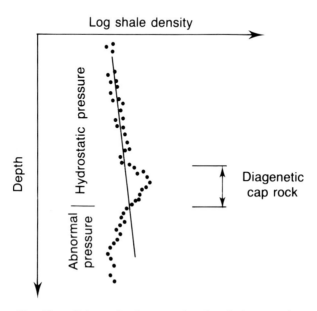

Fig. 40. — Schematic diagram of a chemical cap-rock.

There is much discussion as to the origin of these mineralogical concentrations, but the consensus is that they are of a secondary or diagenetic nature.

Some authors (LOUDEN, 1971) consider that diagenetic cap rocks can be the cause of underlying abnormal pressure, while others (FERTL & TIMKO, 1970 and MAGARA, 1981) believe they are the consequence of it. The authors prefer the latter hypothesis.

The problem is how to account for a zone of preferential carbonate precipitation in the midst of clay series.

The precipitation of cations out of solution in fact depends on variations in pressure, temperature, pH and ionic concentration.

A hydrostatically pressured sequence generally displays only gradual variations in these parameters. On the other hand, where this sequence makes contact with underlying undercompacted formations, parameters change abruptly. This can encourage carbonate precipitation.

It is to be noted that diagenetic cap-rocks do not necessarily accompany undercompacted sequences. They are not present, for instance, in the area of the Niger Delta.

LIST OF METHODS

Predictive methods — regional geology — geophysical methods	Before drilling
Parameters while drilling — Drilling rate — d exponent — sigmalog — normalised drilling rate — M.W.D. (measurements while drilling) — torque — drag	While drilling (real time)
Mud parameters — pit levels — mud flow — pump pressure	While drilling (real time)
— mud gas — mud density — mud temperature	While drilling (not real time)
Cuttings analysis — lithology — shale density — shale factor — shape, size, abundance — cuttings gas	While drilling (not real time)
Wireline logs — resistivity — sonic — density/neutron — gamma ray	After/while drilling
Direct pressure evaluation (formation tests) — drill stem tests — wireline formation tests	After drilling
Well seismic check — checkshot — VSP	After drilling

Even so, if it is accepted that they are formed later than abnormal pressure, they become a factor in maintaining it.

3.3. PREDICTION

3.3.1. *REGIONAL GEOLOGY*

Full appreciation of abnormal pressure problems requires knowledge of more than just the lithostratigraphy of a region. It is also necessary to investigate the hydrodynamics,

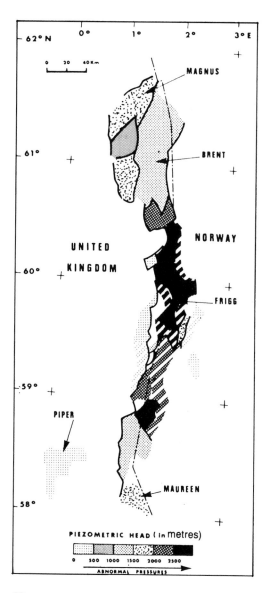

Fig. 41. — Piezometric map of the Jurassic reservoirs in the Viking basin (Chiarelli & Duffaud, 1980, reprinted by permission of the American Association of Petroleum Geologists).

structure, lithology (seal, drainage, sand/clay ratio), clay mineralogy, geothermal gradient, the maturity of its organic matter and so on. If a large amount of data is available, statistical analyses can be carried out with the aim of drawing up maps and overlays for prediction purposes, showing such aspects as lithology, pressure and compaction.

Piezometric maps give an appreciation of abnormal pressure distribution factors. As an example, refer to the piezometric map of the Viking basin (CHIARELLI & DUFFAUD, 1980) which suggests a logical link between abnormal pressure distribution and the geology and tectonics of the region (Fig.41).

Well data can be used to map other features, such as isobaths at the top of the undercompacted zone (Fig.42).

It must always be remembered that using regional maps for making predictions at local level is not without a degree of risk.

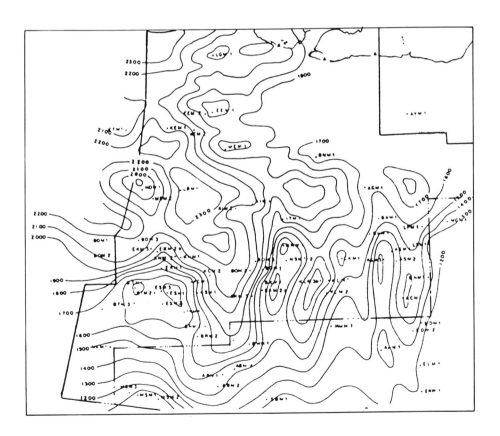

Fig. 42. — Isobaths at the top of an undercompaction zone (Rio del Rey/Cameroon).

In addition to regional maps of pressure distribution (which, incidentally, it is advisable to update in the course of exploring a particular basin), methods are available which attempt to reconstruct the effects of compaction during burial. Mathematical models allow the simulation of sediment burial and permit the prediction of formation pressure. It is thus possible to suggest pre-existing overpressures and their effects on migration.

3.3.2. *GEOPHYSICAL METHODS*

Present day methods of exploiting seismic data can provide numerous clues for detecting abnormally pressured zones. As well as structural information, in certain instances they can provide the following :
- estimated interval velocities,
- the approximate lithology and facies of the sequence,
- direct hydrocarbon detection (bright spots etc),
- detection of abnormal pressure tops and quantitative pressure evaluation,
- high resolution, shallow depth investigation and disclosure of gas pockets and dismigrations (gas chimneys).

A rapid review of seismic methods currently in use along with some methods of interpretation will allow us merely to glimpse the subject. The geologist must be aware of these methods when talking to geophysicists, but their use is normally reserved for the specialist.

3.3.2.1. Seismic methods

Very high resolution seismic

A technique generally used for studying the seabed. It has a resolving power down to less than a metre, and its depth of investigation is limited to 50-100 metres. It is important for platform anchorage, and can also reveal gas pockets and gas chimneys close to the surface.

High resolution seismic

With a resolution in the 1-5 metre range and a depth of investigation reaching between 1 000 and 1 500 metres, this method is an adjunct to conventional seismic methods in the superficial blind spots of the so called "twilight zone". It was developed at the request of the insurance companies after accidents in the Gulf of Mexico attributed to gas pockets, mud volcanoes, faults, abnormally pressured zones, etc.

Conventional seismic exploration

This is the classic technique. It has a lower resolution, in the 5 to 50 metre range, but a depth of investigation extending to several thousand metres. It is the most important source of information about abnormally pressured zones in the vicinity of targets for petroleum exploration.

The seismic section itself conveys a vast amount of information. It reveals gas zones (bright spots), faulting and diapirs. It provides an indication of lithologies and facies, and zones of undercompaction can be detected.

Analysis of interval velocities by deduction from seismic velocities is particularly useful when assessing the development of compaction and the sand-clay ratio. Later we shall examine the use and limitations of this method.

Three-dimensional seismic

Using an acquisition line-spacing of only 50-100 m instead of wide seismic "loops", the 3D method gives a subsurface scan on a regularly spaced grid of points on the XY plane instead of a pattern of lines.

The 3D seismic method thus allows migration of lateral events in all three dimensions establishing the geometry of structures with greater accuracy, and by using "horizon slices" lateral acoustic variations of a given seismic horizon can be defined in 3D.

In offshore 3D seismic this reduced line spacing allows for the simultaneous acquisition of high and very high resolution seismic, thereby avoiding multiple seismic campaigns.

Seismic "S" wave

The foregoing techniques are concerned with primary or compressional seismic waves ("P" waves), in which particles move in the direction of propagation. An "S" wave seismic method is currently being developed involving transverse or shear waves, where particle movement is perpendicular to the direction of propagation. They are also called secondary waves because they are slower than primaries. Since "S" waves do not propagate in liquids, their use is currently limited to onshore applications, although the study of converted "S" waves is a research subject.

Using data on the "P" and "S" waves it is possible to determine the parameters of a rock's elasticity, the Poisson's ratio, modulus of elasticity, and so on. From these it is possible to make predictions, in particular concerning fracture gradients and any possible drilling problems, hole condition, etc.

3.3.2.2. Interpretation

□ *Reflection analysis*

The classic way of representing transit times is by means of a seismic section. Time-related images of echos received for each fixed point on the surface are aligned side by side. Unbroken reflecting features (horizons) reproduce a lithostratigraphic interface. This provides a display of subsurface structures in the vertical plane of the line of acquisition. By examining the continuous nature of reflectors as far as a reference well it is possible to work out correlations for use in defining the predicted geological cross-section.

Sometimes it is also possible to ascertain the different sequences of sedimentation by breaking the image down into sequences of seismic wave trains. This can give useful information about the sedimentation pattern.

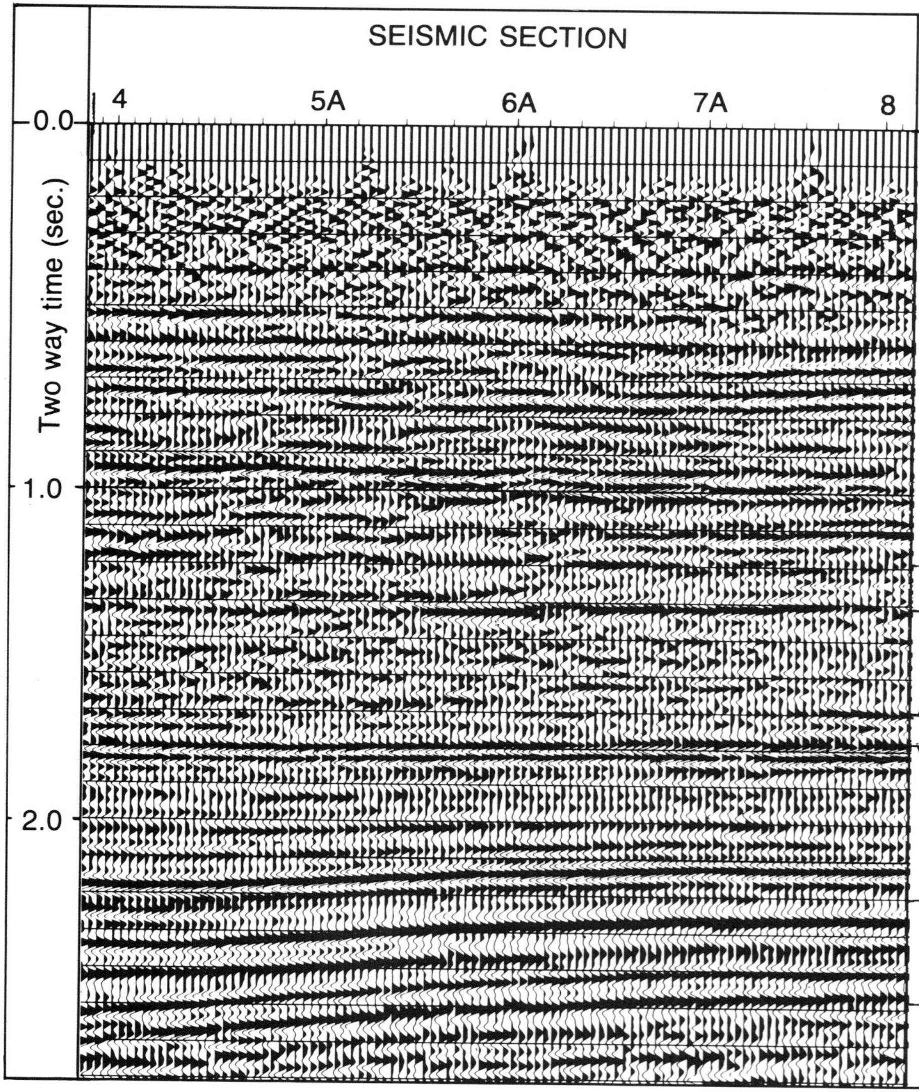

DEPOSITIONAL SETTING	TYPE OF REFLECTION
Deltaic plain	Discontinuous high frequency reflections
Shore - line	Discontinuous reflections
Deltaic plain	Blind zone
Shore - line	Discontinuous reflections Variable amplitude
Littoral deposits (Delta front)	Continuous reflections High frequency Low to medium amplitudes
Transition zone	Continuous reflections Low frequency Medium to high amplitudes
Marine shales	Blind zone

SEISMIC SECTION

Two way time (sec.)

By analysing the subsurface continuity of seismic horizons together with the external shape and internal parameters of reflections (their amplitude, phase and frequency), it may be possible to establish seismic facies corresponding to the depositional setting (Fig.43). By using seismic wave train sequences and facies to identify sedimentation patterns it is possible to arrive at an overall distribution of lithologies. This is called seismic stratigraphy.

Undercompacted zones can be revealed by the nature of low-frequency reflections. If these are poor or reflections are absent (blind zone) this could mean a monotonous sequence of undercompacted clays (Fig.44). The transition zone, limiting the undercompacted series, may sometimes be revealed by a few high amplitude reflections followed by an apparent low frequency wave train.

These criteria for the exploration of undercompacted zones are, however, not conclusive and may be due to other processes such as salt diapirism, a compact uniform series, reefs, laccoliths, etc.

However, if these indications occur together, they strengthen the likelihood of undercompacted clay being present, especially if the correlations or regional geology suggest a comparable interpretation.

It must be emphasised that the time section produced by seismic techniques can be distorted with respect to the true picture at depth if the structure is complex. In such cases, 3D seismic methods can give a more precise picture of deep-lying structure. In such cases computer modelling techniques may be used.

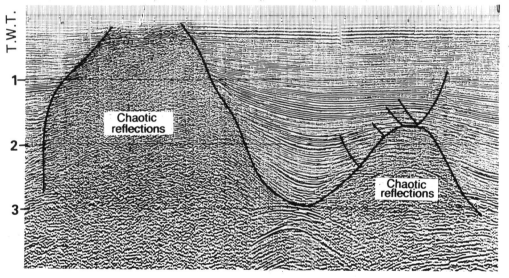

Fig. 44. — Shale diapirism in the Rio del Rey region (Cameroon).

Fig. 43 — Sample interpretation of seismic facies.

In any event, at the wellsite the subsurface geologist can use dipmeter and intermediate sonic data, or even a VSP, to correlate the geological log with the seismic section. When approaching an undercompacted zone or a growth fault, the recognition of already drilled markers enables more accurate estimation of the thickness left to drill before penetrating the risk zone.

□ *Interval velocities*

Where structures are not very complex and the series is sufficiently thick, it is possible to evaluate transit times and calculate the propagation velocity for each interval in the formation.

This velocity is a function not only of the density, porosity and fluid content of the rocks, but also of their elastic properties and stress conditions. The interval velocity alone is insufficient for estimating all these parameters, but if observations recorded from neighbouring wells are taken into account, vertical and lateral variations can be evaluated.

Two aspects of velocity analysis are useful in detecting pressure anomalies :

— establishing velocity/depth curves translated into Δt transit times (ie. pseudo-sonic log). Undercompacted zones, by virtue of their lower density, higher porosity, and abnormally low vertical stresses, have lower velocities.

— the interval velocity which is dependent on the lithology and, for a given lithology, on its state of compaction. For normal compaction conditions, velocity gradually increases with depth.

The velocity of an interval is a function of its maximum burial, but for a tectonically inactive subsident basin, velocity may be linked directly to depth. The curve of normal compaction when velocity is expressed on a logarithmic scale is a straight line, and is known as the compaction trend.

There are several laws defining this relationship, among them the Chiarelli-Serra law :

$$V = Ae^{BZ}$$

or :

$$LogV = A + BZ \text{ on a semi-logarithmic scale.}$$

where V = interval velocity
A and B = constants
       Z = depth

Faust's law introduces the factor of geological age into the formula by postulating that velocity increases uniformly with age. The linearity of this relationship is only valid for a given geological period.

$$LogV = A + B \, LogZ + C \, logT$$

where C = constant
       T = geological time

A basin offshore Western Ireland, where there has been continuous subsidence from the Jurassic to the Miocene, provides an example showing the normal compaction trend and the exponential curve crossing the Faust lines (Fig.45).

74

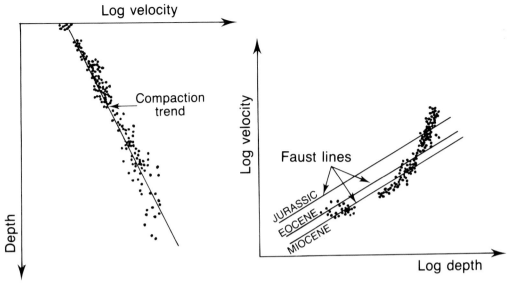

A - Semi - logarithmic plot

B - Log - log faust plot

Fig. 45 A and B. — Velocity/depth curves offshore Western Ireland.

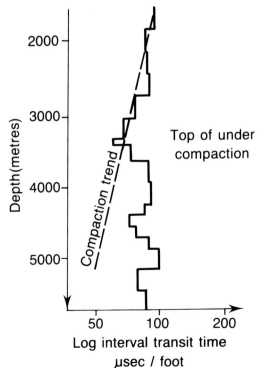

Fig. 46. — Interval velocities reveal an undercompacted zone.

In 1968 PENNEBAKER showed the usefulness of establishing interval transit time/depth curves for deltaic regions.

These curves firstly provide initial depth estimates for tops of undercompacted zones and secondly may allow the evolution of pore pressure to be evaluated.

Figure 46 is an example of a typical curve clearly showing the presence of an undercompacted zone.

Quantitative pressure evaluation may be carried out using either the equivalent depth method or the Eaton method (section 4.1).

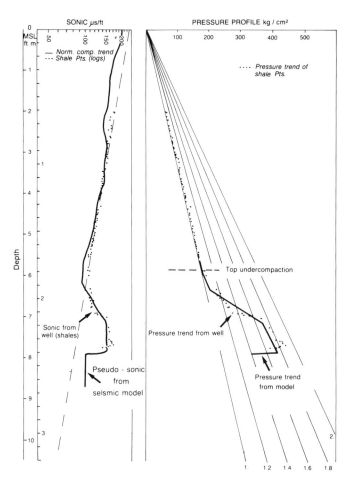

Fig. 47. — Comparison of a pseudo-sonic log and formation pressure evaluated from seismic interval velocities prior to drilling with observed results (Malaysia).

Interval velocities at depth are determined on the basis of estimated lithology. Errors of depth will be smaller the more is known about the lithology (eg. if there are neighbouring wells) and the simpler the underlying structure (eg. a deltaic clay-sand sequence).

The automatic processing of interval velocities, when lithology is sufficiently known, allows the production of a pseudo-sonic log which can then be evaluated in the same way as a sonic log (sections 3.5.1.2 and 4.1). Figure 47 is an example from offshore Malaysia showing not only the excellent correlation between the pseudo-sonic before drilling and the wireline sonic, but also between the formation pressures calculated prior to drilling and those observed.

To convert velocity anomalies into their corresponding pressure anomalies it is necessary to eliminate lithological effects on interval velocities. If there are no reference wells nearby it is advisable during drilling to check the compaction trend, using intermediate sonic logs.

□ *Estimating the sand/shale ratio*

This method is used successfully in deltaic zones such as the Gulf of Mexico and Nigeria. It is based on the fact that, on a semi-logarithmic plot of Δt vs. depth, the points for normally compacted clay are the slowest. A trend line passing through these points represents the clay trend. Another line drawn parallel to it based on a velocity 25 % higher defines the sand trend. The position of the measured velocities in between these trend lines gives an estimate of the sand/shale ratio (Fig.48).

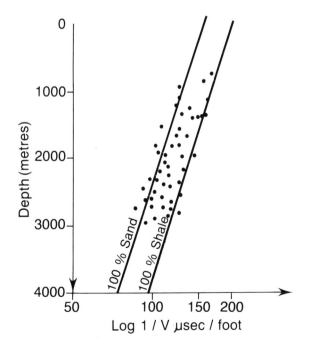

Fig. 48. — Estimating the sand/shale ratio.

We have already seen the significance of the sand/shale ratio value in abnormal pressure distribution (section 2.1), and knowledge of the value of this ratio will lead to a greater capability to predict undercompacted zones.

Reliable interpretation of velocity analyses relies on information about a number of criteria depending on terrain, signal quality and subsequent processing :

— unforeseen changes in lithology
— high angle dip
— faulting
— complex tectonics
— static corrections
— normal move-out corrections
— multiple reflections
— abnormal seismic paths.

□ *Amplitudes*

The amplitude of the signal reflected from the contact between two layers depends on the interface reflection coefficient. This coefficient is a function of the contrast between the acoustic impedances of each layer. Acoustic impedance is the product of the density and the acoustic velocity.

The presence of gas sometimes creates significant amplitude anomalies. Studying such anomalies is the very basis of detecting hydrocarbons directly from seismic data.

The detection of gas pockets prior to drilling is vital to the safety of drilling operations, especially offshore.

On the other hand, lateral amplitude variations can also be due to lateral facies changes which must be taken into account when extrapolating on the basis of reference wells.

3.3.3. *GRAVIMETRY*

This method is generally used to investigate the major structural elements of a basin and the configuration of its underlying basement on a regional scale.

For a given datum point, gravimetry incorporates all density variations occurring in a vertical interval of several thousand metres. Density contrasts in geological formations create gravimetric anomalies. These can arise from a number of causes occurring at different depths, and are therefore complex to interpret. On the other hand, using gravimetry in conjunction with seismic techniques can help resolve uncertainties of either a gravimetric or seismic nature. Indeed, it is essential to use seismic techniques for a detailed interpretation of gravimetric data. Using a time section with identified seismic horizons, it is possible to calculate the gravimetric effects caused by geological formations of known geometry and density. Subtraction reveals gravimetric anomalies associated with geological phenomena which do not show up on seismic sections. This technique has been used for example to determine the presence of salt diapirs in Gabon and the Aquitaine Basin.

Since undercompacted formations can display density contrasts amounting to several tenths $g \cdot cm^{-3}$, it is reasonable to assume that they can be revealed by gravimetric techniques provided the volume of the sediments concerned is big enough.

Onshore gravimetry is usually of a higher resolution than that offshore. In zones of contrasting density where there is potential for petroleum exploration, more attention should be given to acquiring and using gravimetric data which can help to identify zones where the risk of abnormal pressure is high.

3.4. METHODS WHILE DRILLING

3.4.1. *REAL TIME METHODS*

3.4.1.1. Penetration rate

If all else is equal, penetration rate gradually declines as depth increases due to the decreasing porosity caused by the weight of overlying sediments. The method has the potential to detect any significant porosity changes. In fact it has long been known that the penetration rate increases when drilling into undercompacted shales (Figure 49).

In the early sixties, following the penetration rate was the most common method of detecting abnormal pressure. The relative change in penetration rate even provided a first estimate of pressure values.

It has since been found that raw penetration rate includes so many influencing factors that it is too random to be used as a detection method.

The following factors all have a major influence :

— lithology,
— compaction,
— differential pressure,
— weight on bit (WOB),
— rotating speed (RPM),
— torque,
— hydraulics,
— bit type and wear,
— personnel and equipment.

Before taking a closer look at this list, a few comments are in order about how the bit operates downhole at the rock face. The effect this has on the penetration rate will be seen later on.

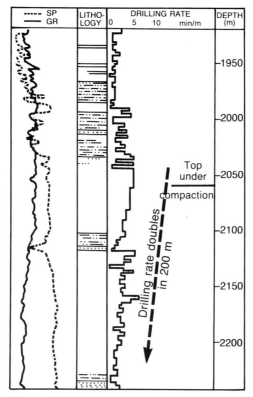

Fig. 49. — Increased penetration rate while drilling undercompacted shales (Odum 1-Nigeria).

□ *Description of the formation breakdown mechanism*

The efficiency of a tooth bit depends on its ability to shatter rock and remove fragments from the bottom of the hole. The process uses the impact of each tooth on the rock face to form a series of small craters.

There are four stages to the process (Fig.50) :

a) Impact

Bit-tooth pressure on the formation increases to the limit of the rock's mechanical strength.

b) Wedge formation

Once the mechanical strength limit has been exceeded, the rock forms a pulverised wedge beneath the tooth. This wedge compacts and horizontal stress develops.

c) Fracture

Horizontal stress increases until the rock fractures and forms a crater.

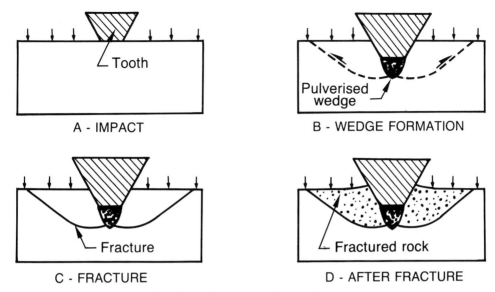

Fig. 50. — Crater formation mechanism.

d) After fracture

The crater consists of fractured rock.

The ease with which fractured rock is removed from the crater depends on the differential pressure at the bit and how much internal friction is present to stop fragments moving along the fracture. In open-air conditions they would actually be ejected with considerable force.

If the mud weight is too high, the increased differential pressure leads to high friction along the fractures, causing fragments to clog.

□ *Lithology*

This is the major factor controlling penetration rate changes. The drillability of a rock depends on its porosity, permeability, hardness, plasticity and abrasiveness, as well as the cohesion of its constituent particles.

All else being equal, a change of penetration rate reflects a change of lithology. If the analysis of cuttings taken at depth is cross-checked against changes in penetration rate, lithological events which typify abnormal pressure can usually be identified. These include cap rocks and monotonous shales. When examining compaction, penetration rate analysis is in two stages. The first stage identifies argillaceous beds and the second examines how penetration rate changes within them.

Small changes in the lithological composition of the shales themselves (clay mineralogy, silt or carbonate content etc.) can significantly alter the rate of penetration. It is common that an increase in silt content can reduce shale drillability up to a certain point, after which drillability improves again. Unlike most of the other parameters, it seems unlikely that such changes in lithological detail will ever be quantified. They depend for their assessment on the experience of the geologist.

□ *Compaction*

The compaction of a sediment is reflected by its porosity, that is to say the extent of matrix grain-to-grain contact.

Given unchanging lithology and no alteration in any of the other variables, penetration rate gradually declines as compaction increases. On the other hand if penetration rate increases in a uniform argillaceous series it reflects undercompaction. The relative change in penetration rate is a function of the degree of undercompaction. It is then possible to correlate the observed compaction state with a comparable state at a shallower depth (section 4.1.1.).

□ *Differential pressure*

Differential pressure (ΔP) is the difference between the pressure exerted by the mud column and the formation pressure (also called the pore pressure). We have already seen the part which ΔP plays in expelling cuttings.

For any given lithology, penetration rate slows as differential pressure increases, and vice versa. For example, a ΔP of 35 kg/cm^2 (500 psi) can cause the rate to slow by around 50 % in comparison with a ΔP of 0 (GOLDSMITH, 1975).

After many laboratory trials it has been possible to draw up typical graphs (Fig.51), and these compare well with one another. In view of the difficulty of simulating pore pressure within a shale, the experiments were conducted using permeable sediments (CUNNINGHAM & EENINK, 1958 ; ECKEL, 1958).

Research based on drilling data confirms that a comparable relationship exists for shales (VIDRINE & BENIT, 1968 ; PRENTICE, 1980) (Fig.52). These graphs show that relationships vary from region to region. In view of these regional variations, similar research needs to be carried out basin by basin if such graphs are to be used for evaluating ΔP.

In undercompacted shales, higher penetration rates are caused both by lower ΔP and increased undercompaction. Put another way, reduced ΔP can come about in two ways. The mud hydrostatic pressure may be reduced or pore pressure may be increased due to a decrease in compaction. In this case it is difficult to distinguish between the effects of these two parameters, especially as undercompaction is not necessarily the only cause of abnormal pressure.

Some authors believe that compaction has a negligible effect, implying that there must be a direct relationship between penetration rate and ΔP. This hypothesis is probably only valid over short intervals.

It is difficult to extend this relationship to include negative ΔP conditions (pore pressure greater than mud column pressure), owing to the scarcity of available data. It is not uncommon to drill shales which create negative ΔP, but permeable layers are so rare that it is no easy matter to measure pressure and obtain accurate information on the ΔP.

82

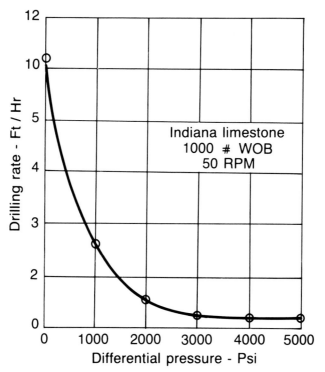

Fig. 51. — The effects of differential pressure on penetration rate (laboratory research by Cunningham & Eenink, 1958, courtesy of SPE of AIME).

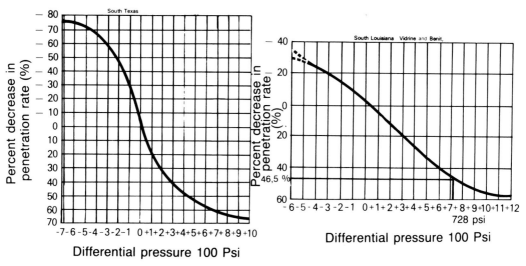

Fig. 52. — The relationship between changes in penetration rate and differential pressure (statistical analysis in South Texas and South Louisiana — Prentice, 1980).

In some cases, penetration rate analysis may enable the definition of an optimum differential pressure which will reduce drilling time without jeopardizing safety standards.

It is tempting to quote GOLDSMITH (1975) on the subject, recalling the simple but sometimes effective rules used by our predecessors : "Many people have learned the hard way that kicks are usually preceded by an increased drilling rate and some have learned that increasing mud weight to keep a "constant" drilling rate will prevent most kicks".

□ **Weight on bit (W.O.B.)**

Changes in weight on bit have more effect on penetration rate than any other drilling parameter.

Generally speaking, penetration rate increases with the weight on bit.

A minimum weight — called the "threshold weight" — is needed to get drilling started. This corresponds to the minimum energy needed for the bit teeth to cause cratering. For example, laboratory analysis of the Berea sandstones (USA) showed that this threshold was 43 tonnes for a tooth with a working face of 1/2 x 1/8" at a ΔP of 21 kg · cm² (300 psi). The threshold weight can be negative in the case of a formation which is only slightly consolidated, since jetting alone is sufficient to ensure penetration.

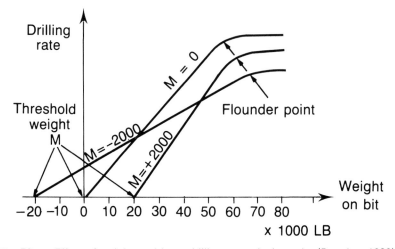

Fig. 53. — Effect of weight on bit on drilling — typical graphs (Prentice, 1980).

Above the threshold weight M, penetration rate rises almost proportionally with bit weight. Above a certain value known as the "flounder point", the drilling rate stops rising since the bit teeth become jammed in the rock. Although the cones increase their rate of contact, they are unable to ensure further penetration. Bit cleaning then becomes less effective (Figure 53).

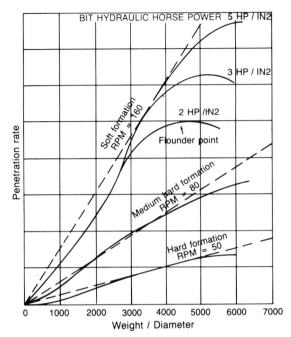

Fig. 54. — Relationship between penetration rate, rotating speed and the weight/diameter ratio for various formation hardnesses (modified from Anadrill document).

The idea of a flounder point is valid only for soft formations (Fig. 54). In this instance, optimum drilling conditions are in the upper section of the graph, where a linear relationship between penetration and weight/diameter ratio does not apply. This ratio more accurately describes how the force is applied at the bit face.

Figure 54 also shows that hydraulics are important for displacing the flounder point to higher drilling rates in soft rocks. This is because bit cleaning is improved by greater hydraulic flow. The chart was plotted on site using optimum rotating speeds for the rock types concerned. Curve shape is only slightly affected by rotary speed.

□ **Rotating speed (RPM)**

It was initially considered that the relationship between penetration rate and rotation speed was linear (BINGHAM, 1964 ; JORDEN & SHIRLEY, 1966).

VIDRINE & BENIT (1968) and also PRENTICE (1980) considered the relationship to be exponential (Fig.55) :

$$R = N^a$$

where R = drilling rate,
       N = rotating speed,
       a = exponent defined empirically on the basis of wellsite tests for a given lithology and weight on bit.

85

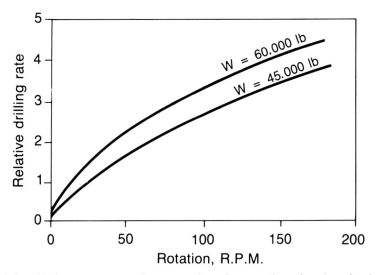

Fig. 55. — Relationship between penetration rate and rotating speed as a function of weight on bit for a given lithology (Prentice, 1980, after Vidrine & Benit, 1968).

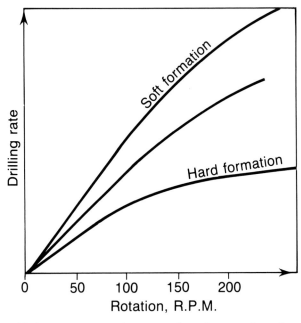

Fig. 56. — Relationship between penetration rate and rotating speed as a function of formation hardness (modified from Zoeller, 1978).

Prentice's graphs have an exponential appearance because the bit teeth gradually spend less time in contact with the formation as rotating speed increases.

Later research has shown that the shape of the curve depends on the lithology (Fig. 56). The relationship in soft formations is nearly linear, but the harder the rock in question, the shorter will be the linear part of the graph. It is likely that the amount of tooth to formation contact time needed to initiate rock breakdown is higher for a hard formation than for a softer one.

□ *Torque*

This parameter is never taken into account directly, since it is very difficult to assess. Surface measurements cannot separate bit torque from string torque. The use of measurements while drilling (M.W.D.) will probably allow the relationship between penetration rate and torque measured at the bit to be established.

Torque at the bit is a measure of the amount of energy needed to break down the rock. This energy is proportional to the product of torque and rotating speed.

It should be noted that an increase in torque at the bit is balanced by a twisting moment from the drill string. The twisting moment corresponds to an angular displacement of the rotary table in relation to the bit. A straight line on the drill string turns into a helix, reflecting the fact that the drill stem has actually shortened.

More often than not, torque is actually stored in the string. This causes weight on bit to fall to a point where the torque is released. This in turn increases weight on bit, and this effect repeats itself at a regular frequency, rather like percussion drilling.

□ *Hydraulics*

The effect of hydraulic flow on penetration rate varies for different degrees of consolidation.

Some authors suggest that there is a linear relationship between hydraulic flow at the bit and drilling rate. But the importance of the effects of hydraulic flow on penetration rate are not fully understood at the moment, and it is not yet possible to draw up a relationship which satisfies every situation.

It is still worth noting that a change in the flow rate can cause a change in the penetration rate.

Mud properties can also affect penetration rate. How they do this is not easy to discover, since many mud characteristics are interdependent.

— *Viscosity*

Effective cleaning of the bit face is particularly dependent on mud viscosity. A low-viscosity, turbulent fluid is more effective than a viscous, laminar one. Low viscosity at the bit may improve penetration.

— *Water-loss (filtration rate)*

It is believed that in some circumstances water-loss can affect penetration rate. This happens as follows : fluid percolates into the fractures caused by the bit teeth and helps to expel rock fragments. This may be mainly because water-loss helps to bring mud pressure and pore pressure into equilibrium.

— *Suspended solids*

Solids can have the effect of reducing immediate water-loss, and in certain circumstances this can limit the penetration rate.

If there are too many solids suspended in the mud, penetration can be impeded because the teeth are prevented from making clean contact with the formation.

This effect is thought to be relatively insignificant.

□ **Bit type and wear**

Optimising the rate of penetration chiefly depends on matching the bit type to the formation.

The usual critical parameters for tri-cone bits are tooth height and spacing, amount of axial offset per cone, and resistance to wear.

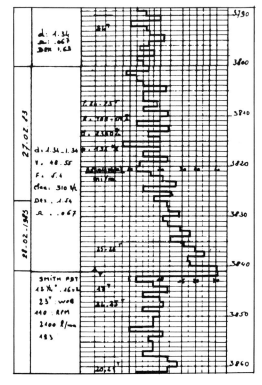

Fig. 57. — The effects of bit wear on penetration rate with unchanging lithology (field document).

Bits are classified by the hardness of the formations they are designed to drill (IADC classification). Selecting the right bit depends on having adequate data about the formations to be drilled, as well as the tools available on site. Variations in lithology sometimes mean that the optimum choice is not possible.

A major change in bit type distorts the value of the drilling rate and alters drilling performance in the event of changes in lithology. This is a hindrance when interpreting progressive changes in the penetration rate.

For these reasons, when approaching undercompacted zones the bit should not be changed to a type other than the one already in use.

Each bit type displays different resistance to wear depending on the nature of the formations being drilled and the drilling parameters. Knowledge of bit wear is important for confirming the lithological interpretation of the rate of penetration (Fig. 57). At the end of its useful life, a bit can mask changes in lithology, compaction or differential pressure due to a decrease in penetration rate under the effects of wear.

Tri-cone bit-wear affects both teeth and bearings. Tooth bits undergo gradual tooth wear, but bearings can wear out quite abruptly once they are no longer water-tight. Insert bits tend not to wear out gradually, but instead their inserts break off in hard, abrasive formations. Insert breakage depends on how well the bit is matched to the formation, on the rotary speed and on vibration.

Later we shall see how, in certain circumstances, drilling rate can be corrected to allow for bit wear.

A diamond bit proceeds by making scratches or grooves, not by cratering. Relationships between penetration rate and drilling parameters follow different rules. Rotating speed and possibly hydraulic flow are the main factors, and their relationship with the drilling rate is linear.

□ *Personnel and equipment*

Appropriately chosen drilling parameters regularly applied to the lithology under investigation help to optimise penetration rate. An experienced driller can make a vital contribution in this area. Rig equipment (eg. drive-gear capacity) can impose an upper limit on parameters (RPM etc.).

CONCLUSION

Under ideal drilling conditions in shales, penetration rate can be thought of as dependent on porosity, and therefore a way of detecting undercompaction. In normal use, however, many parameters affect the reliability of the measurement. To use it properly we have to employ drilling models, such as the "d" exponent, the sigmalog or the normalised drilling rate.

3.4.1.2. "d" exponent

Various ways of "normalising" penetration rate have been formulated over the last twenty years. Their aim is to eliminate the effects of drilling parameter variations and arrive at a representative measure of formation drillability. Field work has shown that, short of using a computer, the solution known as the "d exponent" method is the simplest and most reliable. This technique was formulated in the Gulf Coast shales and takes major variables into account. It has proved to be so successful that it is still, with some refinements, the most frequently used method.

BINGHAM (1964) suggested the following relationship between drilling rate, weight on bit, rotating speed and bit diameter :

$$\frac{R}{N} = a \left(\frac{W}{D}\right)^d \qquad (3.1)$$

where
- R = drilling rate in feet per minute
- N = rotating speed in revolutions per minute
- W = weight on bit in pounds
- D = bit diameter in inches
- "a" = lithological constant
- "d" = compaction exponent (dimensionless)

JORDEN & SHIRLEY (1966) solved this equation for the "d" exponent by introducing constants which would allow standard petroleum industry units of measurement to be used and give values of "d" which could be handled easily. Since constant "a" is not required where the lithology is constant, a = 1 is assumed by definition.

Standard US units (3.2)

$$« d » = \frac{\log_{10}\dfrac{R}{60\,N}}{\log_{10}\dfrac{12\,W}{10^6\,D}}$$

- R = feet/hour
- N = revolutions/minute
- W = pounds
- D = inches

Standard "Metric" units (3.3)

$$« d » = \frac{1.26 - \log_{10}\dfrac{R}{N}}{1.58 - \log_{10}\dfrac{W}{D}}$$

- R = metres/hour
- N = revolutions/minute
- W = tonnes
- D = inches

Ratio R/60 N in formula (3.2.) is always less than 1. It represents penetration in feet per drilling table revolution.

Where lithology is constant, the "d" exponent gives a good indication of the following :
— the state of compaction (ie. porosity)
— differential pressure

Calculating "d" exponent in shales makes it possible to follow their stages of compaction and reveal any undercompaction (Fig.58).

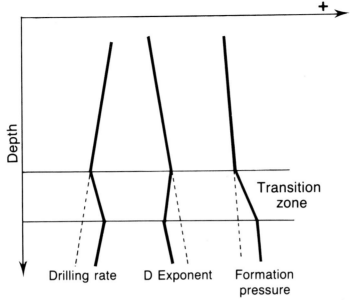

Fig. 58. — Schematic diagram of "d" exponent in an undercompacted zone.

Any decrease in "d" exponent when drilling an argillaceous sequence is a function of the degree of undercompaction and also of the value of the associated abnormal pressure.

□ *Corrected "d" exponent (dc)*

Differential pressure is dependent on pore pressure and mud weight. If the mud weight is changed, "d" exponent will change. "d" must therefore be corrected for changes in mud density so that it properly represents the differential between formation pressure and regional hydrostatic pressure.

REHM & McCLENDON (1971) suggested the following correction :

$$"dc = "d" \times \frac{d1}{d2}$$

where "dc" = corrected "d" exponent
 "d" = "d" exponent
 d1 = formation fluid density for the hydrostatic gradient in the region (1.00 to 1.08)
 d2 = mud weight (d_{eqv})

Figure 59 shows how an increase in mud weight can mask or attenuate changes in "d" exponent while drilling an undercompacted zone.

Before taking a closer look at the methods of application, we shall first examine :

— the validity of "d" and "dc" formulae,
— the effect of parameters not incorporated in "d" and "dc".

91

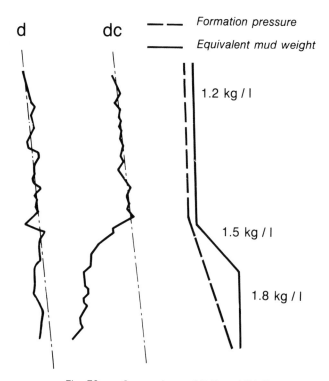

Fig. 59. — Comparison of "d" and "dc"

□ *Criticism of "d" and "dc" formulae*

The "d" exponent formula is empirical, especially as far as correcting for mud weight is concerned. Experience shows that such corrections are generally effective. However, in cases where ΔP is very high the correction required is so large that "dc" drops to excessively low values which then show little variation. In this case "dc" is not practical to use.

Parameters not included in "d" and "dc"

Compaction and ΔP parameters are not included because it is precisely these, or variations in them, that we wish to highlight.

Other parameters not taken into account are as follows :

— lithology : "d" exponent can only be applied to the same lithology (clays and shales for the detection of undercompacted zones).

— mud hydraulics : this parameter has a more marked effect in unconsolidated sequences, where jetting is important. In consolidated formations, the efficiency of bit-face cleaning definitely has an effect where there are significant variations in hydraulic flow at the bit, but the implications are not fully understood (see section 3.4.1.8).

— bit type and wear : a major change of bit type is likely to cause a shift in the trend of "d" exponent values. This effect is easily identified, and just as easily overcome during interpretation as we shall see in the next two sections.

Gradual bit wear causes a gradual rise in "d" exponent values. This is superimposed on the compaction effect. If bit wear occurs abruptly at the end of bit life, the effect is easier to spot but can mask entry into an undercompacted zone.

□ *Ways of correcting for tooth wear*

There are several ways of trying to correct for the effects of bit wear on "d" exponent. They usually derive from two different approaches based on an inversely proportional relationship between penetration rate and a function of tooth wear as follows :

$$R = \frac{1}{F(H)} \qquad (3.4)$$

where : R = relative penetration rate = R_1/R_0
with :
R_1 = rate of penetration corresponding to bit wear H
R_0 = rate of penetration corresponding to zero bit wear
$F(H)$ = bit wear function
H = tooth wear (on a scale of 0 to 1) ; bit wear is usually noted on a scale of T1 to T8 at the wellsite — eg. T4 = 4/8 = 0.5

giving : $R_0 = R_1 \times F(H)$ $\qquad (3.5)$

The value of R obtained by these methods is substituted in place of the unadjusted value of R in equations (3.2) and (3.3).

The function F(H) differs in the two methods as follows :

GALLE & WOODS *correction method (1963)*

The formula for calculating bit wear function F(H) expresses an almost linear relationship between bit wear and F(H) (Fig.60a).

$$F(H) = \sqrt{0.93\ H^2 + 6\ H + 1}$$

giving : $R_0 = R_1 \sqrt{0.93\ H^2 + 6\ H + 1}$

VIDRINE & BENIT *correction method (1968)*

Tests on argillaceous formations have empirically established a linear relationship between wear and F(H) (Fig.60a).

$$F(H) = 1 + 2.5\ H$$

giving : $R_0 = R_1 (1 + 2.5\ H)$

Figure 60b shows the slight difference between the results of the two methods of correction.

For example, wear of 0.5 (T4) gives the following reductions in penetration rate :

— Galle & Woods : 51 %
— Vidrine & Benit : 44 %

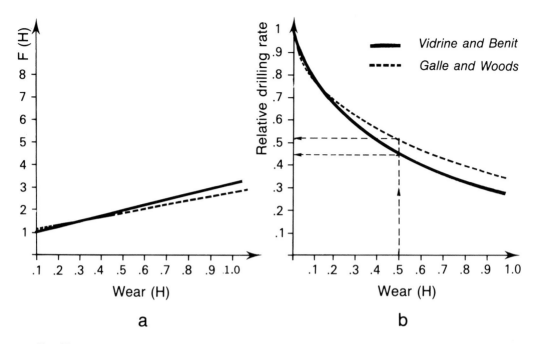

Fig. 60. — Relationships expressing wear/wear function and wear/relative penetration rate

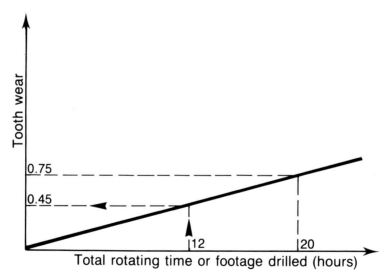

Fig. 61. — Example of a bit wear trend.

These methods have only limited use, however, because the degree of bit wear cannot be known with certainty while drilling is in progress. Either of the following relationships can be used instead (Fig.61) :

- a linear relationship between wear and total rotating time or between wear and revolutions per minute,
- a linear relationship between wear and footage drilled.

Bit wear trends are established for each bit. Wear is then estimated while drilling, by reference to the trend for the previous bit used in comparable circumstances.

Example (Fig.61) :

- previous bit T 6 : H_o = 0.75
- total rotating time : t_o = 20 hours
- current bit : rotating time : t = 12 hours

Calculating current bit wear (can be read off directly from the graph) :

$$H = H_o \cdot \frac{t}{t_o}$$

$$H = 0.75 \times \frac{12}{20} = 0.45$$

The same procedure can be used for calculating wear on the basis of footage drilled.

The bit wear function F(H) was established for tooth bits. For other types of bit, some authors suggested applying a correction coefficient K to F(H) :

- tooth bits : K = 1
- insert bits : 0.4 < K < 0.6
- diamond bit : 0 < K < 0.2
- no bit wear correction : K = 0

Corrected "d" exponent formula taking bit wear into account :

$$dcs = \frac{d_1}{d_2} \cdot \frac{1.26 - \log \dfrac{R\ F(H)^K}{N}}{1.58 - \log \dfrac{W}{D}}$$

Bit wear corrections are frequently used by mud logging companies, but are not entirely satisfactory for the following reasons :

● Any relationship between penetration rate and wear is not realistic. Given bit wear of 25 %, the relationship suggests that the penetration rate is reduced by over a third in comparison with a new bit ! Such a reduction is obviously excessive. Bits pulled out of hole with 50 % wear are commonly seen at the end of their useful life still drilling at 85 % of their initial drilling rate (GOLDSMITH, 1975).

These relationships were developed close on twenty years ago, at a time when tooth bits were used in all circumstances. With developments in technology, tooth bits are currently used only in relatively soft formations. As we have seen, these are the very conditions under which the relationship introduces too large a correction.

Furthermore it is unsatisfactory to introduce correction coefficients for other bit types using a relationship which is based on the wear characteristics of tooth bits, because the wear processes involved are quite different.

● Bit wear correction formulae do not take lithology into account. In particular they ignore the hardness and abrasiveness of the formation being drilled. In the simplest case of uniform lithology, there is some justification for calculating bit wear from a linear relationship between wear and drilling time. On the other hand, where there are interbedded formations which differ markedly in character, it is unrealistic to establish such a relationship. Any attempt at using a correction coefficient for bit wear to allow for abrasiveness is doomed to failure because of the obvious difficulties in drawing up a scale to quantify this parameter.

● Bit wear evaluation while drilling takes no account of weight on bit. To restrict wear calculation to a function of the number of times the teeth strike the working face, by concentrating on drilling time or the total number of revolutions, is unsatisfactory. The energy expended on each impact also has an appreciable effect on the rate of wear.

To sum up, penetration rate corrections which are currently in use to allow for bit wear are unsatisfactory, and must be used with caution. Valid corrections can however be established regionally within a well defined set of conditions, but only provided there is sufficient data for a statistical analysis to be carried out.

□ *Recommendations for the use of the "d" exponent*

a) *Equipment*

● sensors :

— penetration rate
— weight on bit
— rotating speed
— mud weight in and out

● calculation methods

— the nomogram in Fig.62 : often used in the past before the advent of data processing methods. Inconvenient to use, its inaccuracies are often accentuated by distortion related to the way it is printed.
— "d" exponent slide rule : calibrated in standard US units. Produced by IMCO in the early seventies to overcome the disadvantages of the nomogram and improve the speed and accuracy of calculations.
— programmable calculator : satisfactory calculation speed. Data has to be selected and checked by the user and entered by hand. Inexpensive, portable and easy to use. Data can be thoroughly checked during use, and therefore particularly suitable for monitoring the "d" exponent at the wellsite.

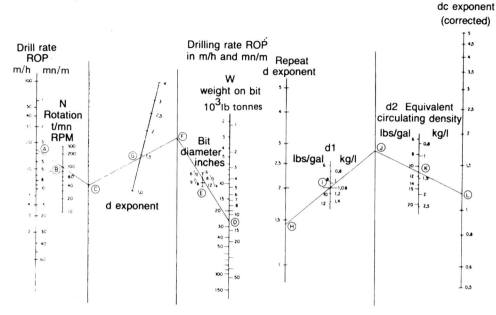

Fig. 62. — Nomogram for calculating "d" exponent.

— microcomputer : equipment supplied on request by all mud-logging contractors. Uses data from on-line sensors to carry out automatic calculations and plot graphs. Because these operations are automatic there are no sorting and validation routines. Post-calculation checks therefore need to be made, particularly to ensure that lithology has been taken into account.

b) *Data selection*

— drilling rate, weight on bit, rotating speed : instantaneous values for these parameters should not be used. Instead use the average value for each parameter based on homogeneous intervals.

Where there is an on-line mud logging unit systematically calculating the value of "d" exponent every metre or half metre, mean values are in theory calculated automatically, but it is worthwhile checking. If data are entered manually, they can be averaged visually on the mud logging charts. *On no account should averages be based on data from different lithologies.*

— mud weight : in general, use average mud-weight out referring to the same interval as the other parameters.

The use of equivalent circulating density (section 1.2) is recommended. Measurements while drilling (M.W.D.) provide more precise data both for E.C.D. and weight on bit, but are rarely available.

97

— regional hydrostatic gradient (d1) : usually between 1.00 (Nigeria) and 1.07 (Gulf Coast). The value is established from test data taken in hydrostatic series. In the case of an exploration well with no reference data, use an average value (such as 1.05) and apply corrections following RFT or tests. Any errors introduced into the calculations will be small to insignificant.

— lithology : as shales are the only lithology whose compaction state can be followed by the "d" exponent, extra care must be taken in correctly logging samples, so that the chosen drilling parameters will correspond to shale beds. Changes in shale lithology are likely to alter penetration rate, and need to be noted if "d" exponent is to be interpreted correctly. Attention must be paid to factors such as consolidation (eg. washout proportions, plasticity and induration), silt or carbonate content, and accompanying minerals such as pyrites.

c) *Calculating and displaying "d" exponent*

● Formula :

It is preferable to use the Jorden & Shirley formula corrected for mud weight (p. 90).

● Corrections :

If detailed statistics on bit performance for an explored region are not available, it is preferable to make no corrections at all than use bit wear corrections suggested by the service companies.

In the past an empirical correction was often applied in respect of bit type. Insert or diamond bits were considered to perform less well than tooth bits, so the method was to take an inch off the diameter used in the calculation. As bits of this type have improved in performance, there is no longer any justification for the correction, and in fact it should be actively discouraged.

Note that if the bit most suited to the formation is selected, bit changes will have little or no effect on "d" exponent.

● Choice of intervals

In the main, where calculations are done manually the intervals are selected before the calculation is carried out. The choice depends on lithology, and only shale intervals are used. These are identified by the penetration rate and validated later by cuttings analysis.

If calculations are carried out automatically, intervals are chosen afterwards.

Some authors suggest that it is also possible to use "d" exponent to define how compaction changes with depth in sands and carbonates. If grain size is constant, it is possible to estimate the compaction regime for uncemented sands. But variations in both porosity and grain size affect penetration rate, and mask alterations in compaction. Texture and cementation vary so widely in sandstones and carbonates that they do not reflect the effects of compaction adequately, and are generally poor indicators for investigating the subject. On the other hand it may be possible to assess the porosity of such formations by calculating the "d" exponent.

If there are no thick argillaceous series, there is sometimes a need to obtain "d" exponent values for thin shale beds. In such cases, computer analysis is likely to incorporate values external to the bed concerned. This can be overcome manually. For instance, in predominantly sandy sequences such as the Benin formation of the Niger Delta, it is essential to establish "d" exponent values for all the minor interbedded shales in order to establish the compaction trend.

● Calculation frequency :

Since computerised mud-logging units carry out calculations at preset intervals, we are concerned here only with manual calculations. These must be carried out as often as possible, depending on the penetration rate. Five-metre intervals are suitable unless there is interbedding, in which case all argillaceous levels need to be listed. This becomes all the more important the less frequently they occur. When approaching the top of a prognosed undercompacted zone or when there is any doubt, the interval may be reduced to every one or two metres.

In order to establish an accurate compaction trend, the plot should be started as soon as possible, that is to say when the clays have become sufficiently indurated to minimise the effects of jetting (Fig.63). In offshore situations it is not often possible to calculate meaningful values for "d" exponent before reaching a depth of 500 to 800 m.

● Graphical representation :

The plot should preferably be on tracing paper so that interpretation attempts can be carried out on disposable blueprints.

The horizontal axis ("d" exponent) must be logarithmic and the vertical axis (depth) linear.

For an effective display of how compaction changes with depth, it is essential to use a reduced depth scale. 1 :2 000 is adequate for day-to-day interpretation and 1 :5 000 for the final document. A scale of 1 :10 000 may be suitable for comparative documents correlating "d" exponent with other parameters (Fig.64).

When plotting points, it is better not to join them up. There is no lithological justification for doing so, and it can lead to misinterpretation.

It is easier to interpret the "d" exponent if the document includes related information such as the lithological column, casing details, bit changes, mud weight, etc. (Fig.63).

d) *Interpreting the "d" exponent*

● Plotting the compaction trend

Where there is a significant thickness of homogeneous shale, plotting the compaction trend usually presents no major problem. Preference is given to pure shale points. These

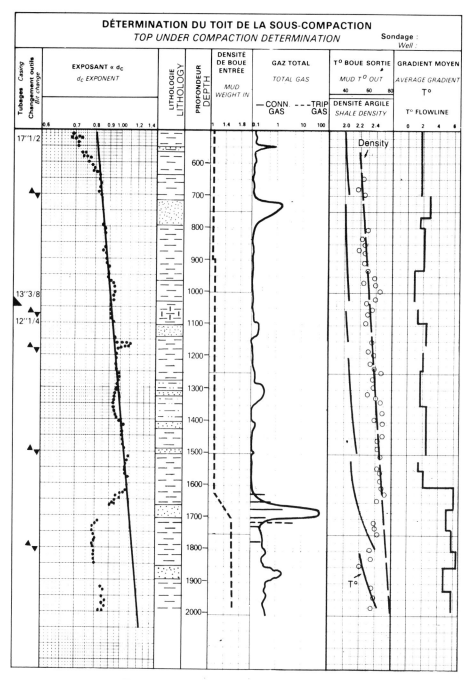

Fig. 63. — Standard wellsite compaction log.

100

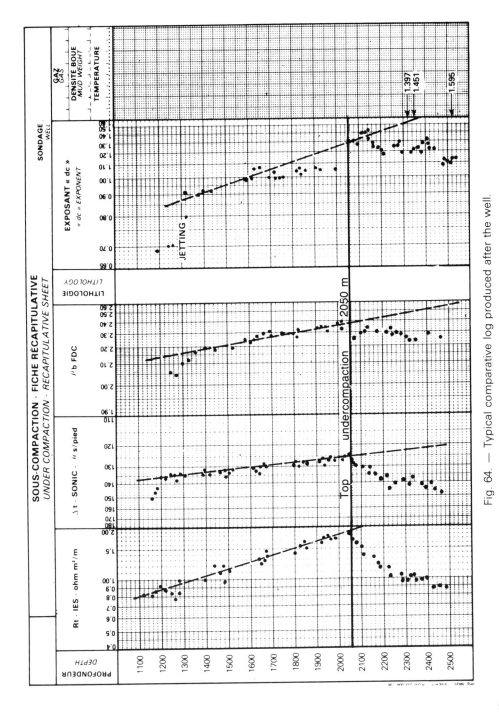

Fig. 64. — Typical comparative log produced after the well.

101

appear to the right of the scatter. It is necessary to exclude lithological or drilling anomalies such as bit balling.

Figure 63 is a specimen compaction trend plot exhibiting the following common anomalies :

— from 500 to 640 m : low "d" exponent (jetting effect)
— from 960 to 1 040 m : "d" exponent too high (calcareous shale)
— from 1 160 to 1 180 m : "d" exponent too high (bit balling)
— from 1 280 to 1 410 m : "d" exponent too low (shale with silty laminations)

If there are no pure shales, plotting the trend is a more difficult matter. The points on the compaction trend must refer to comparable types of shale. For instance, the compaction trend for a silty shale sequence is to the left of the trend line for pure shales, but the trend for calcareous shales is to the right.

Positive sedimentary sequences (increase in detrital content with depth) are frequently observed in deltaic zones and may be wrongly interpreted as a transition zone. This explains why lithological observations are so important.

Even if a compaction trend is not plotted from pure shales, it may still be possible to determine whether there is any undercompaction. Pressure values can be accurately assessed from the "d" exponent if the shale is of the same type throughout.

● Regional compaction trend

Compaction effects show up as a positive slope ("d" exponent increases with depth) in the compaction trend. A vertical trend line indicates incomplete compaction. It is a good idea to establish regional slopes for "d" exponent in normally compacted series, provided this statistical analysis is confined to a homogeneous lithological body and comparable drilling conditions. Only the slope is to be taken into account, as absolute values may differ for identical depths. Such investigations have given satisfactory results, notably in the Gulf Coast, where there is an abundance of statistical data. A study involving a limited number of wells in Angola established that a regional slope of 0.33/1 000 m applies in the Tertiary sequences of the North-West offshore only.

● Compaction trend shifts

Once the dominant trend for a given shale has been established, there are no lithological reasons to shift it unless the nature of the reference shale changes over a large depth interval.

The effect of bit wear on the "d" exponent, which is often seen at the end of bit life as excessive values, does not require a shift (Fig.65a). Notice that this effect can make it difficult to identify penetration into the undercompacted zone (Fig.65b). It is advisable that when approaching such a zone, the bit should not be pushed to the limit of its useful life, but used if possible with optimum drilling parameters.

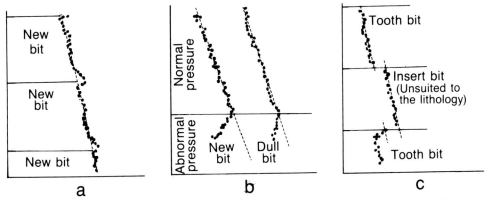

Fig. 65. — The effect of bit type and wear on "d" exponent (Courtesy of EXLOG).

Situations requiring a shift in the compaction trend

— Change of bit type :

If a bit type is used which is unsuited to the formation, the trend has to be temporarily shifted towards higher values (Fig.65c). Unfortunately the most suitable bits are not always available at the wellsite and so such shifts are not uncommon.

— Change of diameter :

Changing the diameter and significantly altering drilling parameters can make it necessary to shift the trend (Fig.66a).

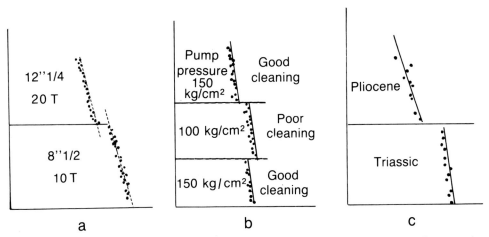

Fig. 66. — Situations requiring shifts in the compaction trend : a : change of drilling diameter ; b : change of mud pressure ; c : change of geological series (Courtesy of EXLOG).

— change of drilling parameters :

Since the empirical formula for "d" exponent only partially corrects for changes to drilling parameters, these need to be optimised. Poorly adapted values or significant variations can lead to a shift in the value of the "d" exponent.

— change of hydraulic flow :

The "d" exponent is calculated on the assumption that the bit face is being cleaned efficiently. Major changes to hydraulic flow at the bit will affect the result of the calculation (Fig.66b).

— unconformity :

Formations of very different ages and states of compaction may require both a shift of trend line and a change of slope (Fig.66c). At great depth there is almost negligible reduction in porosity, and the compaction trend becomes nearly vertical.

In the example of Fig.66c, the depth at which the Triassic is encountered is clearly shallower than its maximum burial.

— directional wells :

When a directional well is being drilled, the weight recorded at the surface is greater than the actual weight being applied at the bit due to string friction. The true weight on bit is governed by the hole angle and the nature of the bottomhole assembly. Thus "d" exponent values will be higher than their true value unless true weight on bit values are available from M.W.D. measurements.

In addition note that the compaction trend must be established using vertical depths and not drilled depths.

Shifting the trend line

The technique for shifting the trend assumes identical compaction on both sides of the change. The first few points after a bit change should be ignored as they cover the running-in period for the new bit. In any event, the shift must give a new trend line with the same slope as its predecessor (except in the case of major unconformities or faults).

□ **Conclusion**

Experience shows that the "d" exponent is an efficient technique. If used within its intended limits under appropriate drilling conditions, it makes a decidedly valid contribution to detection of undercompacted zones. It should however be associated with other methods of detection, even when being used under ideal conditions.

Although data processing certainly helps to simplify some routines, it is also true that if the operator evaluates the data and chooses the points, control is improved and interpretation made easier. The human factor is essential !

3.4.1.3. Sigmalog

The Sigmalog was developed in the Po Valley in the mid-seventies as a joint venture between AGIP and Geoservices. The aim was to solve the shortcomings of the "d" exponent while drilling overpressured sequences of carbonates, marls and silty shales in deep wells.

Since Sigmalog entails more complex calculations than the "d" exponent, the technique cannot be used without the aid of a computer.

The Sigmalog is the variation with depth of the Sigma Factor, also called "total rock strength". The Sigma Factor takes the same factors into account as the "d" exponent.

The initial relationship is as follows :

$$\sqrt{\sigma_t} = \frac{W^{0.5} \cdot N^{0.25}}{D \cdot R^{0.25}}$$

where $\sqrt{\sigma_t}$ = raw sigma or total rock strength (dimensionless)
$\quad$ W = weight on bit (tonnes)
$\quad$ N = rotating speed (r.p.m.)
$\quad$ D = bit diameter (inches)
$\quad$ R = penetration rate (metres/hour)

To correct for mud weight $\sqrt{\sigma_o}$ is calculated such that :

$$\sqrt{\sigma_o} = F \cdot \sqrt{\sigma_t}$$

where $\sqrt{\sigma_o}$ = corrected sigma, or "rock strength parameter"

$$F = 1 + \frac{1 - \sqrt{(1 + n^2)} \, \Delta P^2}{n \, \Delta P}$$

where ΔP = differential pressure of mud to formation fluid corresponding to the normal hydrostatic gradient (kg $\cdot$ cm^2)
$\quad$ n = factor expressing the time required for the internal pressure of cuttings not yet cleared from the bit face to reach mud pressure.

n was established for the Po Plain on the basis of $\sqrt{\sigma_{t'}}$:

$$\sqrt{\sigma_{t'}} = \sqrt{\sigma_t} + 0.028 \left(7 - \frac{Z}{1\,000}\right)$$

where $\quad$ Z = depth in metres
$\quad \sqrt{\sigma_{t'}}$ = correction factor for determining pore pressure at shallow depths.

If $\sqrt{\sigma_{t'}} \leqslant 1 \quad n = \dfrac{3.25}{640 \sqrt{\sigma_{t'}}}$

If $\sqrt{\sigma_{t'}} > 1 \quad n = \dfrac{1}{640} \left(4 - \dfrac{0.75}{\sqrt{\sigma_{t'}}}\right)$

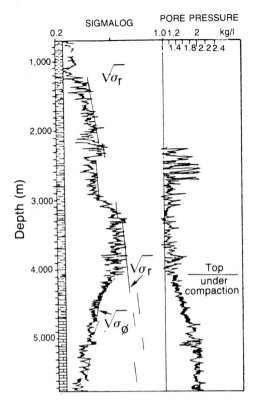

Fig. 67. — Specimen Sigmalog — Italy (Bellotti & Giacca, 1978).

The value of n is a function of formation permeability and porosity. As a general rule, $\sqrt{\sigma_r} < 1$ for sands and > 1 for shales. This means that n is greater for shales than for sands and reflects the fact that the bit face is more difficult to clean in shales.

Changes in n have a minor effect on the calculation of $\sqrt{\sigma_o}$. It is therefore not a problem to apply it to other sectors than the Po Valley where it was defined, as the error introduced will also be minor.

Rock strength parameter $(\sqrt{\sigma_o})$ is plotted against linear horizontal and vertical scales (Fig.67). For identical lithologies Sigmalog behaves like compaction, ie. increases with depth.

The highest values represent the lowest porosity. The normal compaction trend passes through these points.

The slope of the trend usually remains constant at 0.0881/1 000 m.

The trend is defined as follows :

$$\sqrt{\sigma_r} = \alpha \, \frac{Z}{1\,000} + \beta$$

where $\sqrt{\sigma_r}$ = parameter defining the reference compaction trend
α = trend slope
β = intersection of the trend with the horizontal axis for Z = 0
Z = depth in metres

A shift is required every time there is a change of lithology, diameter or bit type, but the slope stays the same. If values of $\sqrt{\sigma_o}$ start to fall without any change of lithology or drilling conditions, this suggests an increase in porosity and/or formation pressure (Fig.68). Sigmalog has been suggested as a method for determining porosity (BELLOTTI & GERARD, 1976).

The empirical formula used is as follows :

$$\phi = \frac{1}{1{,}4 + (9\sqrt{\sigma_o} \cdot \sqrt{\sigma_r} / \sqrt{\sigma_\phi})}$$

where ϕ = porosity
$\sqrt{\sigma_\phi}$ = trend of the $\sqrt{\sigma_o}$ points most to the right (Fig.68).

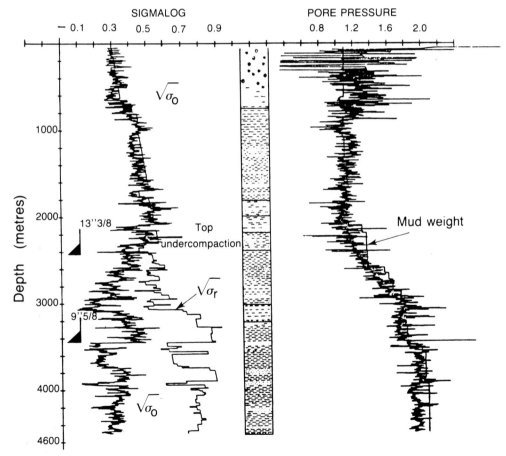

Fig. 68. — Automatic Sigmalog processing (Egypt) (Cesaroni et al., 1982).

This empirical relationship was developed in the Po Valley and must be used with caution in other regions.

Automatic Sigmalog processing is illustrated in Figure 68 below.

Pore pressure is evaluated from differences between the reference trend $\sqrt{\sigma_r}$ and $\sqrt{\sigma_\phi}$ (Fig.67). The $\sqrt{\sigma_r}$ trend must be shifted to allow for changes of formation, bit or diameter, such that :

$$\frac{\sqrt{\sigma_{r1}}}{\sqrt{\sigma_{r2}}} = \frac{\sqrt{\sigma_{\phi1}}}{\sqrt{\sigma_{\phi2}}}$$

According to its authors, Sigmalog is a better method for normalising drilling parameters, apparently because anomalies resulting from their variations are attenuated. But there are some examples from other parts of Italy which show that "d" exponent can in fact be less affected than Sigmalog by variations of this nature (see casing points and lithological variations in Fig.69). The horizontal scale chosen for both tools is fundamental to this comparison.

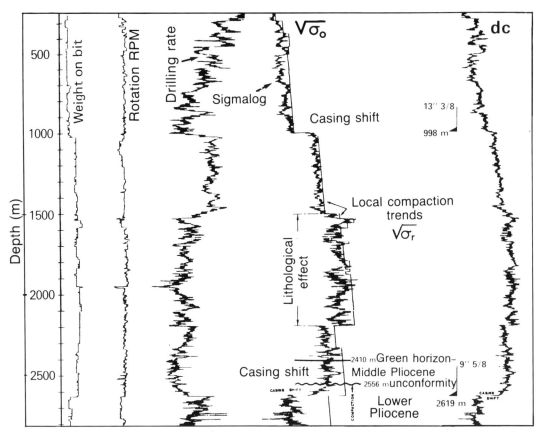

Fig. 69. — Comparison between curves for Sigmalog and corrected "d" exponent (Italy).

The efficiency and limitations of Sigmalog are very similar to those of "d" exponent. This is especially true with regard to lithology, as the method must be restricted to argillaceous sequences.

CONCLUSION

Sigmalog may be used in place of "d" exponent. But the method is not easy to use and therefore ill suited to unexplored basins. The limitations of Sigmalog are the same as for "d" exponent. In particular its use should be entirely restricted to clays and shales. On the other hand, Sigmalog relies too heavily on the operator's judgment when determining the various trend shifts required. At the same time the interpreter has little control over the calculation stage.

3.4.1.4. Other drilling rate normalisations

A number of other formulae attempt to improve on the "d" exponent by incorporating parameters which it omits. We shall examine just two of these, as they do appear to have a certain amount of potential. We shall then take a brief look at some of the normalisation methods which mud logging contractors have put forward.

□ *Combs formula*

In 1968 Combs proposed a mathematical drilling model which took into account not only the parameters used by "d" exponent, but also hydraulics and bit wear.

The constants in the following formula were established by regression analysis on the basis of data from just six wells :

$$R = R_o \left(\frac{W}{D \cdot 3500} \right)^{a_w} \left(\frac{N}{200} \right)^{a_n} \left(\frac{Q}{D \cdot D_n \cdot 3} \right)^{a_q} f(\Delta P) \cdot f(H)$$

R_o = drilling rate (ft · hr^{-1})
W = weight on bit (pounds)
D = bit diameter (inches)
N = rotating speed (r.p.m.)
Q = flow (gal · min^{-1})
D_n = jet diameter (1/32")
a_w, a_n, a_q = weight, rotation and flow exponents
$f(\Delta P)$ = differential pressure function
$f(H)$ = bit wear function.

R is defined as the penetration rate for a new bit operating under the following conditions :

— differential pressure = 0
— weight on bit = 3 500 pounds

— rotating speed = 200 r.p.m.

— jet flow = 3 gal/min/1/32 inch

Exponents a_w, a_n and a_q are constants :

$a_w = 1$

$a_n = 0.6$

$a_q = 0.3$

Correction for bit wear (H) is usually carried out via the following general formula :

$$H = \sum_{outil} \left(\frac{\Delta Z_i}{R_i}\right) \left(\frac{N_i}{200}\right)$$

where R_i and N_i correspond to penetration rate and rotating speed in shale for the interval ΔZ_i. R_i is introduced into the bit-wear calculation in order to make an allowance for formation abrasiveness, though the method of doing so may be rather simplified.

Combs regards the relationship between H and drilling rate as linear, and the function f(H) as constant.

Anyone intending to apply this method to an unexplored basin would be well advised to carry out a similar statistical analysis to re-evaluate the various constants in Combs' formula.

□ *Normalised drilling rate*

Normalised drilling rate was introduced by PRENTICE (1980) on the basis of research by VIDRINE & BENIT (1968). It has a fresh approach compared with other methods.

This model represents the drilling rate which would have been obtained if actual values of the hydraulic and mechanical variables used had been equal to an arbitrary set of "standard values".

$$NDR = R \frac{(W_n - M)}{(W - M)} \cdot \frac{(N_n)^\lambda}{(N)^\lambda} \cdot \frac{Q_n}{Q} \cdot \frac{BPL_n}{BPL}$$

where

NDR = normalised drilling rate

R = raw drilling rate

W = weight on bit

N = rotation

Q = mud flow rate

BPL = bit pressure losses

M and λ = dimensionless parameters

W_n, N_n, Q_n, BPL_n are the predefined standard parameters for each bit.

Corrections for the mechanical parameters of weight and rotation are calibrated by a drilling test (Five Spot Drill-off Test). Such tests must be carried out for each diameter and each bit type (IADC classification). M and λ are determined with different values of weight and rotating speed respectively (Fig.70). These tests are carried out over intervals of 3 to 5 metres where bit wear, lithology and pore pressure are assumed not to alter.

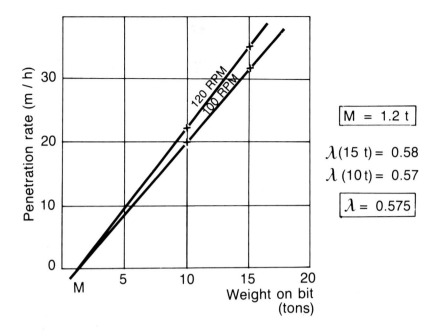

$M = 1.2\ t$

$\lambda(15\ t) = 0.58$

$\lambda(10\ t) = 0.57$

$\lambda = 0.575$

DEPTH m	DRILLING RATE m / h	ROTATION RPM	WEIGHT ON BIT t
1020 - 1023	20	100	10
1023 - 1025	31.5	100	15
1025 - 1028	35	120	15
1028 - 1030	22.5	120	10
1030 - 1032	18	100	10

$$\lambda = \log\ (R1\ /\ R2)\ /\ \log\ (N1\ /\ N2)$$

Fig. 70. — Example of M and λ determination by drill-off test.

If a bit has been drilling for a considerable stretch, it is advisable to redetermine M and λ to check whether these parameters are affected by bit wear or compaction. Any significant drift in these values would call the validity of the model in question.

Bit pressure losses are derived using the following equation :

$$BPL = \frac{K \times Q^2 \times d}{(D1^2 + D2^2 + D3^2)^2}$$

D1, D2, D3 = diameters of jets 1, 2, 3.
       d = mud weight
      K = constant

111

The formula for normalised drilling rate therefore becomes :

$$NDR = R \times \frac{(W_n - M)}{(W - M)} \times \frac{(N_n)^\lambda}{(N)^\lambda} \times \frac{(Q_n)^3}{(Q)^3} \times \frac{d_n}{d} \times \frac{(D1^2 + D2^2 + D3^2)^2}{(D1_n^2 + D2_n^2 + D3_n^2)^2}$$

When values for λ and M have been determined for each bit type, the standard parameters $(W_n, N_n, Q_n, D1_n, D2_n, D3_n)$ are the ones used during the drilling test.

The rate of penetration is also a function of the following factors :
— lithology,
— compaction of the matrix,
— bit efficiency,
— differential pressure at the bit.

To eliminate lithological effects, penetration rates in any formation other than shales are ignored.

The way in which NDR changes with depth Z has been likened by Vidrine and Benit to a linear decrease as long as lithology, pore pressure and mud weight remain the same. This means there are just two phenomena responsible for the decrease : increasing compaction of the matrix and decreasing bit efficiency due to wear.

Once the bit dulling trend has been defined (Fig.71), any deviation of NDR from this trend is interpreted as a variation in differential pressure $\Delta P = P_h - P_p$ at the bit.

Trends are plotted for a given differential pressure for each new bit run (ie. when there is nil wear and a different IADC bit type classification). Each trend has an associated term "dt" ("trend density"). This value, which is expressed in equivalent mud weight, represents the formation pressure and corresponds to the normal variation of NDR represented by the trend. The term dt can be established with the aid of pressure calculations carried out prior to a trend change. Otherwise, pressure measurements can be used, or any other type of formation pressure data such as shut-in drill-pipe pressure, etc. We shall examine the method of calculating pore pressure in section 4.1.5.2.

The trend slope must be established in a normally compacted zone. The slope is then amended in accordance with the IADC classification for the bit concerned in the following way :

— if the new bit is more suited to softer formations than the standard bit, the trend slope is corrected by multiplying penetration rate by $(0.88)^x$, where x is equal to the number of lines separating the two bit types in the IADC classification table.
— if on the other hand the new bit is more suited to harder formations than the standard bit, multiply by $(1.12)^x$ to modify the trend slope.

The new trend line origin is established from the first few values obtained for normalised penetration.

Figure 71 is a plot of normalised drilling rate showing shifts in the trend due to changes of bit type and mud weight.

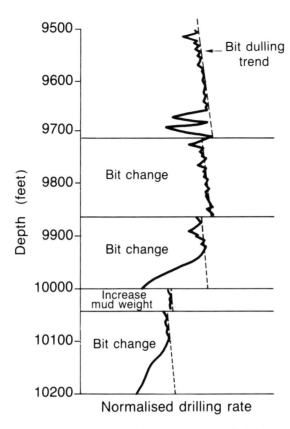

Fig. 71. — Example plot of normalised drilling rate (modified from Prentice, 1980).

In principle, normalised drilling rate has certain advantages over "d" exponent. Analysing penetration rate for each bit on the basis of drilling tests is an original and promising approach which has recently been tested on Elf Aquitaine wells.

Based on data and critical analysis following these tests, the following observations can be made :

— the hydraulic correction is too great. Altering flow by 20 % changes the value of normalised drilling rate by a factor of 2.

— test drilling does not always give a correct definition of M and λ. It also prolongs drilling time.

— Vidrine & Benit's formulae concerning the relationships between drilling parameters and penetration rate are not entirely satisfactory (see section 3.4.1.1.).

— compaction changes are not normalised in this model. According to Prentice, the main influence on penetration rate is variation in differential pressure. In any case, it is not advisable for charts drawn up in the Gulf Coast (Fig.52) to be used just as they stand in other regions.

— the positioning of the normal pressure trend from the first few points is crucial. The frequently scattered data points and the empirical corrections for bit type are both obstacles to trend establishment.

— the method is complex. It needs a microcomputer and highly qualified staff. Even then interpretation is a tricky and somewhat lengthy process.

To conclude, in its present form Normalised Drilling Rate does not seem to be a sufficiently reliable method. It does reflect a need to improve the "d" exponent by a less empirical approach. Although current models are not entirely satisfactory, it should be possible to enhance them by carrying out local statistical studies in greater detail. The problems involved in applying Normalised Drilling Rate emphasise the obstacles facing any attempt to improve "d" exponent by introducing ill-justified and experimentally unproven corrections, as these could well falsify the interpretation of results.

□ **"Drilling Model" — "Nx" (Exlog)**

In 1971, REHM & McCLENDON defined a relationship between "d" exponent and differential pressure (section 3.4.1.2.). In 1973, BOURGOYNE & YOUNG reported their work on establishing a bit-wear model. Exploration Logging drew on both these sets of findings to devise an empirical model called "Nx", based on the drilling parameters, as follows :

$$R = a \left(\frac{N}{100}\right)^{\beta} \cdot \left(\frac{W}{\beta}\right)^{\frac{N_x \cdot d_1}{d_2}} \cdot Te \, (h)$$

where   a = "effective RPM" equating to Bingham's lithological constant (formula 3.1)
           β = RPM exponent
    Te(h) = tooth wear (0 to 1)

It is no easy matter for the operator to determine Nx, as it requires five parameters to be established : a, β and three wear coefficients.

It was so complex that Exploration Logging has recently conducted research into a theoretical "drilling model". It is intended to be operated in dynamic mode, and be capable of identifying and interpreting drilling events.

The model has been available, in principle, since mid-1985, and runs on an IBM/XT. Though it uses the latest advances in computing techniques it is based on relatively old concepts, from such authors as OUTMANS (1959) and GEERTSMA (1961).

Only time and experience will tell whether it has merit, especially as the products proposed (from a porosity log to a pseudo-sonic) are largely outside the usual run of services provided by mud logging contractors.

□ "LNDR" (Baroid)

Baroid's "Log Normalised Drill Rate" is a model based on work by COMBS (1968) and BOURGOYNE & YOUNG (1973). It also incorporates some elements of "d" exponent and Sigmalog.

The method takes the effects of the following parameters into account :

— penetration rate
— weight on bit
— rotating speed
— bit tooth wear
— hydraulic characteristics
— compaction effects
— differential pressure

There are two formulae for deriving LNDR. The first takes account of the drilling parameters, hydraulics and bit wear (3.6). The other takes compaction, differential pressure and matrix stress into account (3.7).

$$LNDR = \log \left[\frac{Re^{1.1H}}{WN^{.6} J^{.15}} \right] \tag{3.6}$$

where
- $LNDR$ = "log normalized drill rate"
- R = penetration rate (ft $\cdot$ hr^{-1})
- W = f(weight on bit) : klbs/4 x bit diameter (teeth)
- N = f(rotating speed) : RPM/100
- J = f(jet impact) : lbs/10 x bit surface
- H = f(bit wear) : 0 to 1
- e = natural log base

$$LNDR = a_0 - a_1 D - a_2 D (d - p) + \frac{0.357 (p-p_n)}{16.4 - (p-p_n)} \tag{3.7}$$

where
- D = vertical depth, ft/1 000
- d = equivalent mud weight, lb/gal
- p = formation pressure, lb/gal
- p_n = normal pore pressure, lb/gal
- a_0, a_1, a_2 = constants : 3.0, 0.1 and 0.156 respectively.

Combining both these formulae gives pore pressure :

$$MNLDR = a_0 - a_1 D + (p-p_n) \left[a_2 D + \frac{0.357}{16.4 - (p-p_n)} \right] \tag{3.8}$$

It should be noted that Baroid uses empirical relationships and constants based on work carried out on a regional scale (Gulf Coast) and as such LNDR cannot be used in other areas without risk.

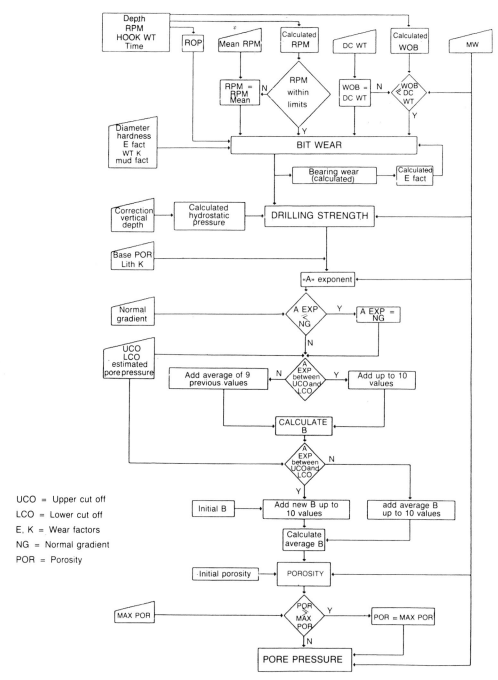

UCO = Upper cut off
LCO = Lower cut off
E, K = Wear factors
NG = Normal gradient
POR = Porosity

Fig. 72. — Flowchart for the IDEL program (Courtesy of Anadrill).

□ IDEL — "A" exponent (Anadrill)

IDEL or "Instantaneous Drilling Evaluation Log" is a computerised method for automatically calculating pore pressure. The principle of the method is to establish a relationship between the energy expended by the bit in drilling a predetermined section of rock, and the lithology, porosity and pore pressure. The equivalent energy, known as the "rock strength", is calculated from the drilling and hydraulics parameters in conjunction with mud and formation characteristics.

"Drill Strength" is used both to establish "A exponent" from which pore pressure is deduced, and to calculate porosity. Figure 72 shows the flowchart for the calculation program.

The originality of this method lies in its ability to differentiate automatically between deviations of A exponent due to lithological effects and those due to the effects of pore pressure (Fig.73a).

Unlike the other methods, this one does not assume in advance that the pressure trend is linear, but plots it by averaging the immediately preceding values. Cut-offs are predefined on the basis of preceding penetration rate. Values which diverge from these are interpreted as lithological in origin (Fig.73b). A significant number of the constants in the "Drilling

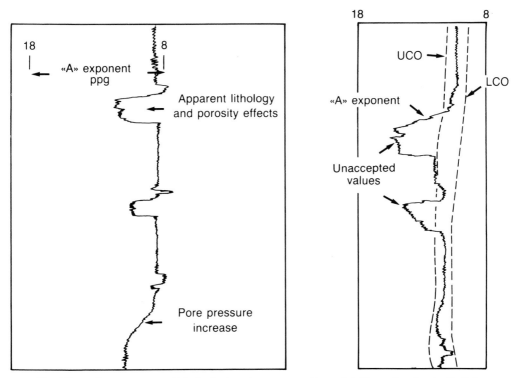

Fig. 73. — A exponent — Lithological and pressure effects (a). The role of cut-offs (b).

Strength" formula were established in the Gulf Coast, and this method must therefore be used with caution in other areas.

3.4.1.5. Torque

As we have already seen, surface measurements of torque integrate both torque at bit and string friction against the borehole walls. As depth increases, so does the amount of contact between the borehole walls and the drill string, so that torque gradually increases too. An anomalous rise in torque can have a number of causes. One of these can be a change in differential pressure associated with entering an abnormally pressured zone.

If there is negative differential pressure (mud weight too low), the mechanical behaviour of the shales may cause torque to rise in either of two ways (note that both processes may occur simultaneously) :
— by swelling of plastic clays, causing a decrease in hole diameter.
— by an accumulation of large cuttings around the bit or stabilizers.

The plastic state of the clays in superficial formations may cause the bit to ball-up. If balling-up occurs at greater depths, it may indicate that the bit is entering a transition zone. Balling-up is usually indicated by a reduced and steady torque.

Although torque is not easy to interpret in view of the many phenomena which can affect it (hole geometry — deviation — bottomhole assembly make-up), it must be thought of as a second-order parameter for diagnosing abnormal pressure.

3.4.1.6. Overpull and drag

Pinching of the borehole walls may cause overpull (hook weight higher than free string weight) while pulling out of hole, or additional weight may have to be applied while going in hole, even perhaps to the extent of re-drilling.

Although such phenomena may be due to abnormal pressure, they may have other causes such as a dog-leg, well deviation, partial sticking due to differential pressure, clay swelling and so on.

3.4.1.7. Hole fill

After tripping or during connections, cavings may have settled preventing the bit from reaching bottom. Wall instability in an area of abnormal pressure may cause sloughing. It should be noted that such fill can have other causes, such as :
— wall instability for geomechanical reasons (fracture zones).
— inefficient well-cleaning by the drilling mud.
— rheological properties of the mud insufficient to keep cuttings in suspension.
— caving of normally compacted clays and shales : unadapted mud (thermal or ionic imbalances).

Information on torque, drill-string drag or hole fill is episodic and details must be noted in a comprehensive document at the wellsite (Fig.63) for crosschecking with the interpretation established from the main parameters.

3.4.1.8. Pit level — differential flow — pump pressure

Measuring pit levels, differential flow and pump pressure provides a means of recognising kicks arising from negative differential pressure. This may be caused by a rise in formation pressure or a decrease in equivalent mud density (losses, mud weight decrease, gas, swabbing etc).

A kick is a critical state of well imbalance, and must be detected as early as possible.

Differential mud flow measurement with the aid of electromagnetic flow meters is currently the best method for early detection of kicks or mud losses. It has the advantage over pit-level measurement that surface mud movements do not have to be accounted for. It also gives a quicker and more accurate response. Because it is difficult and costly to install

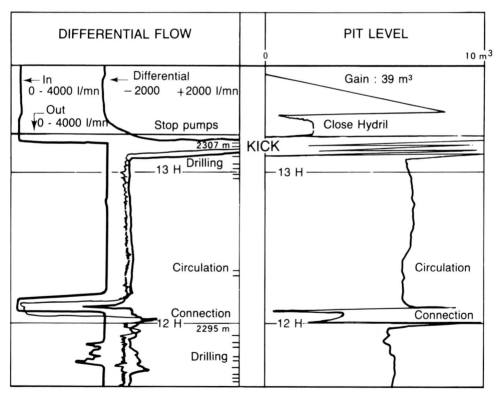

Fig. 74. — Example of a kick in an abnormally pressured sequence (Niger Delta).
The above figure shows an abrupt kick encountered in an undercompacted shale-sandstone sequence.

it is not an automatic choice, even though the alternative flow measurement of pump strokes (in) and paddle (out) does not give very reliable results.

In regions where abnormal pressures are known to exist but where no transition zone can be observed the choice of electromagnetic flow meters is essential, as no other early warning of well imbalance may be available and the sooner a kick is detected the easier it will be to control.

Pump pressure may vary with the lithology of the formations penetrated, especially while drilling with a diamond bit. Such changes can also indicate a kick.

Although a kick is a threat to well safety, it does give valuable information on the formation pressure. Mud weight can be adjusted accordingly, and the interpretation of detection methods such as "d" exponent can be recalibrated.

3.4.1.9. **M.W.D.**

"Measurements While Drilling" were originally designed for continuous measurement of well deviation. M.W.D techniques now provide a range of methods which are significantly

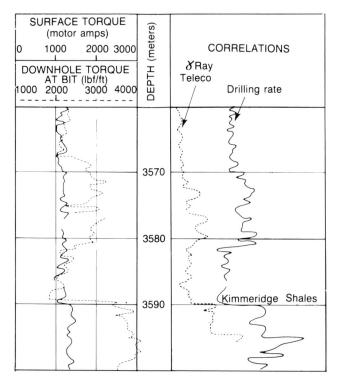

Fig. 75. — North Sea Kimmeridgian shales revealed by torque at bit.

improving the state of knowledge on bottomhole drilling parameters and formation evaluation :

— bottomhole weight on bit,
— torque at bit,
— mud pressure,
— mud temperature,
— mud resistivity,
— formation resistivity,
— formation radioactivity.

If the true weight on bit is known, drilling rate can be normalised with better accuracy.

Fig.75 illustrates the greater sensitivity of the downhole torque measurement which acts as a technique for lithological evaluation. There is a very close correlation between torque, penetration rate and gamma ray measurement. Entry into the Kimmeridgian shales can clearly be seen.

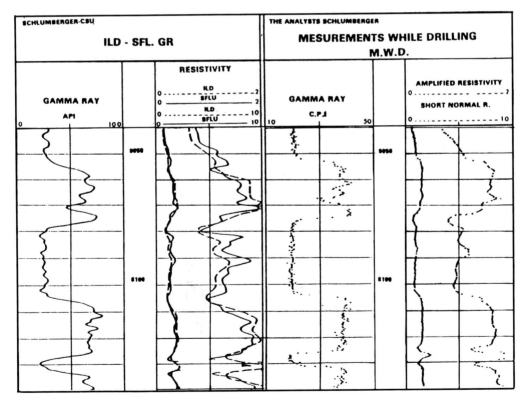

Fig. 76. — Comparison between wireline logging and M.W.D. (TC3) (Gulf Coast) (Courtesy of Anadrill).

Information on true bottomhole mud pressure gives a more accurate view of the effects of swab, surge and annular pressure loss.

Figure 76 shows the very close correlation between data obtained using classic wireline logging techniques and M.W.D. These results were obtained using the TC3 (Anadrill) system and M.W.D. curves are now continuous.

Although downhole mud temperature can provide additional information, it is not easy to apply because it is dependent on so many variables (section 3.4.2.3.).

On the other hand, differential resistivity between mud in the drill pipe and in the annular space may well be considered a kick indicator (section 3.4.2.4.).

3.4.2. *METHODS DEPENDING ON THE LAGTIME*

3.4.2.1. Mud gas

The monitoring and interpretation of gas data are fundamental to detecting abnormally pressured zones. Originally introduced to avoid gas cut mud becoming a threat to the safety of drilling operations, gas detection and analysis now provide the wellsite geologist with information concerning source rock, reservoir and well equilibrium.

Figure 77 summarizes the various possible sources of gas shows.

Gas shows can be categorized according to source as follows :

— cuttings gas : gas released from the drilled formation and by the cuttings moving up the annulus.
— produced gas : gas issuing from the borehole walls. This may be due to caving or swelling, and can also arise from diffusion or effusion if differential pressure is negative.
— recycled gas : if the mud is not completely degassed at the surface, it may be returned down-hole still gas cut.
— contamination gas : from petroleum products in the mud or from thermal break-down of additives. Breakdown of organic matter in shales or thermal effects produced by the bit can also give rise to gaseous hydrocarbons.

The quantity of gas measured at the surface depends on a number of factors :

— the hydrocarbon content of the formation drilled,
— the petrophysical characteristics of the rock (porosity, permeability),
— differential pressure,
— the volume of rock drilled (diameter, drilling rate),
— mud flow,
— mud characteristics (type, viscosity, temperature, hydrocarbon solubility, etc.),
— the measuring equipment (degassing efficiency, instrument precision, etc.).

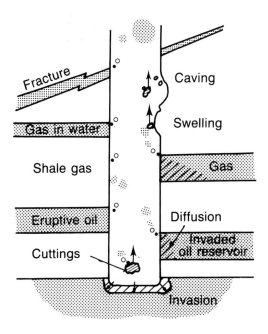

Fig. 77. — The various sources of gas shows.

□ **Background gas**

Background gas is the gas released by the formation while drilling. It usually consists of a low but steady level of gas in the mud which may or may not be interrupted by higher levels resulting from the drilling of a hydrocarbon-bearing zone or from pipe movement such as trips and connections.

An increase in the level of background gas from that found in overlying normally compacted shales often occurs when drilling undercompacted formations. Such an increase comes about for the following reasons :

— a generally higher gas content,
— an increased drilling rate,
— a drop in differential pressure.

The determining factor is the ΔP : if the mud weight is too high it can mask or even eliminate gas shows.

When the well is close to balance ($\Delta P \neq 0$), analysing changes in the level of background gas in shales can often identify when an undercompacted zone has been entered.

Figure 78 shows the close correlation which exists between increases in background gas and drilling rate. When gas volume is corrected for flow and drilled rock volume

123

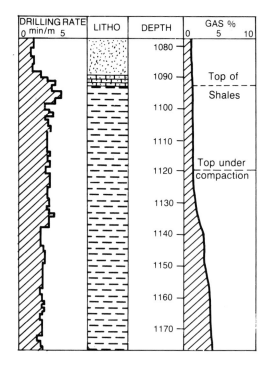

Fig. 78. — Example of an increase in background gas when drilling an undercompacted zone (Niger Delta).

(corrected gas index), it is found in this instance that an increase in drilling rate is not the only cause of the rise in background gas.

If background-gas variations are observed in an argillaceous series while mud weight and drilling parameters remain unaltered, this often indicates that formation pressure has changed.

Other detection parameters must be examined before a decision is taken to increase mud weight.

Any mud weight increase must be gradual. If background gas diminishes while this is being done, it proves that the cause was differential pressure. Adjusting mud density in line with background gas therefore enables well balance to be monitored.

However, if ΔP is negative it can cause instability in the borehole wall and lead to an abnormal increase in the volume of cuttings through caving. Cavings contribute to a rise in background gas which an increase in mud weight is not always enough to absorb. Under these conditions it is advisable not simply to increase mud density, but also to adjust mud characteristics.

It should be noted that when a hole is drilled, rock is removed and its place is taken by a fluid. This leads to an anomalous redistribution of overburden stresses. If the borehole walls become destabilised, changes in mud hydrostatic pressure will have only a slight effect on the new stress distribution.

Background gas is often a good method for detecting and monitoring abnormal pressure. But some undercompacted shales have no gas at all, in which case this parameter cannot be used.

□ *Gas shows*

If porous, permeable formations containing gas are penetrated while drilling, gas shows can occur. Their volume is governed by differential pressure.

Figure 79 shows three different situations which can arise while drilling through the same homogeneous reservoir.

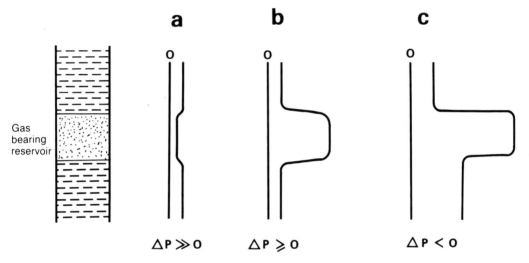

Fig. 79. — The effect of ΔP on gas shows.

a) If mud weight is too high and fluid loss is not checked, gas shows will be reduced as the gas is flushed ahead of the bit.
b) Normal drilling conditions : the gas show exceeds the level of the background gas. Background gas level is identical before and after the reservoir.
c) If differential pressure is negative, the gas show is larger. Gas continues to flow from the reservoir as drilling continues in the non-reservoir below, and this raises the background gas level.

Observing the form and abundance of gas shows can make it easier to detect a state of negative differential pressure. This is very important for the detection of abnormal pressures where no transition zone exists.

125

The presence of connection gas (CG) or trip gas (TG) may be typical of well imbalance. The equivalent density applied to the formation with pumps stopped (static) is lower than the equivalent circulating density (dynamic). When the well is close to balance, the drop in pressure while static may allow gas to flow from the formation into the well.

The quantity of gas observed at the surface when circulation is resumed depends mainly on the following criteria :

— differential pressure,

— formation permeability,

— nature of the gas contained in the drilled formation,

— length of time pumps were halted,

— movement of the drill pipe (swabbing upwards and surging downwards).

Observing the frequency and progression of connection gases can be a valuable aid to evaluating differential pressure. Figure 80 refers to the same case as Figure 78. This is

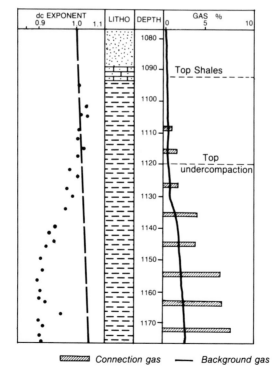

Fig. 80. — Comparison of progressive changes in connection gas - background gas — "d" exponent (Niger Delta).

126

a good example of connection gases occurring more frequently when an undercompacted zone of uniform shales is drilled without increasing mud weight. Note that the connection gas value must be read above the background gas level (eg. if the total CG value is 3 % whilst BG is 1 % the CG recorded is $3 - 1 = 2$ %).

Background gas and connection gas are shown separately, but superimposed to simplify interpretation.

To monitor connection gas correctly the following criteria must be observed :

— lithology : as far as possible, preferential attention must be paid to connection gases from argillaceous sequences. Permeability is then less critical, and the gas arises from diffusion or caving. The contribution of caving to total connection gas should be assessed by observation of the volume of cavings produced.

— connection gases may be compared with one another, provided connection times are fairly uniform. On the other hand, in the case of trip gas (ie. after bit tripping, deviation measurements, etc.) stopping times vary and comparisons are more difficult.

— coming out of hole can produce a temporary condition of negative differential pressure or exaggerate one which already exists. As pipe is pulled, mud tends to rise with it by virtue of its own viscosity, lowering effective mud pressure at the borehole wall. The extent of the pressure drop is governed by the pipe pulling speed, the annulus diameter and the mud rheology. M.W.D. trials have shown that the drop in pressure due to swabbing is generally greater than calculations suggest. ROBINSON *et al.* (1980) report M.W.D. data (Exxon cable telemetry) which indicate pressure drops corresponding to reductions in equivalent density of as much as $0.10 \text{ g} \cdot \text{cm}^{-3}$. Cuttings can accumulate, especially on the stabilisers, and increase the swabbing effect. This reduces equivalent mud weight still further.

In order to keep the effects of swabbing on connection gas to a minimum it is recommended that pulling speed should be kept steady.

When making a connection while approaching an undercompacted zone, the best procedure is to pull the drill pipe slowly, maintaining circulation. Pump stopping times must be as consistent as possible.

As with background gas, if mud weight is too great it can eliminate connection gas, thereby masking a first order detection parameter. Such gas indicators are easier to interpret if the well is close to balance, but it must be kept in mind that there is a risk associated with drilling while connection gas is on the increase. In situations of this kind mud weight must be adjusted. If mud weight remains steady, the following situations may be encountered (Fig.81) :

— *background gas stable — connection gas sporadic* (Fig.81a) : this situation is not characteristic of formation pressure variation. Connection gas may be present due to swabbing, lithological changes or caving. However, variable connection conditions can give rise to this situation in a transition zone. Interpreting this situation is ambiguous.

— *background gas stable — connection gas increasing* (Fig.81b) : typical of entering a transition zone. The stable background gas suggests positive differential pressure,

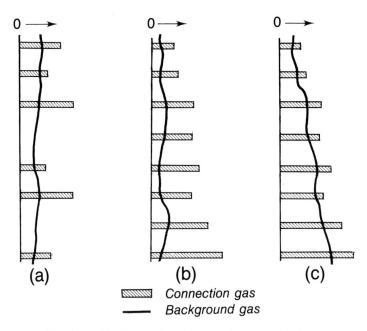

Fig. 81. — Background and connection gas variations.

but the increasing incidence of connection gas reflects a decline in differential pressure.

— *background gas and connection gas on the increase* (Fig.81c) : drilling is procee-ding in negative ΔP conditions or approaching imbalance.

The best information concerning well equilibrium is to be obtained from overall trends in connection gas irrespective of short-term fluctuations. It is generally necessary to wait as long as 40 or 50 metres to define an increasing trend. In fact connection gas is more a method for monitoring developments in pressure than a means of precisely defining the top of undercompaction.

Abnormal pressure is confirmed if, by adjusting the mud weight, the volume of connection gas is reduced. In ideal situations, the way in which connection gas responds to changes in mud density within an argillaceous setting provides continuous and accurate analysis of changes in formation pressure as they develop.

□ *"Normalised" connection gas*

In order to obtain standardised gas data, some companies recommend deliberately creating standard gas shows using a rule known as "10-10-10". The method involves inducing gas slugs under three different sets of equivalent density conditions. Gas shows can then be interpreted more accurately.

There are four stages in the procedure :

i)   drilling stopped, bit on bottom
     continue rotating and
     maintain circulation for   10 minutes
ii) drilling stopped, bit on bottom
     continue rotating with
     circulation stopped for   10 minutes
iii) pull pipe 10 m at preset standard speed
     continue rotating with
     circulation stopped for   10 minutes
iv) one complete circulation while drilling

The following results are obtained :

i)   background gas under equivalent circulating density
     conditions
ii) gas show under static equivalent density conditions
iii) gas show while swabbing
iv) gas transfer to surface.

The method is good in principle, but is time-consuming when applied regularly and may also lead to stuck pipe. One may also question the wisdom of consciously introducing large gas slugs which may lead to well imbalance.

It should however be possible to use variants of the method to reduce the amount of lost time, without sacrificing the principle. Periods could be reduced to five minutes, stage iii (swabbing) could be eliminated, etc.

□ *Gas composition*

The occurrence or increased incidence of heavier gas components is commonly observed when drilling into transition zones. This can be used as a means of detecting undercompacted zones.

Undercompacted clays are often source rocks. If volatile hydrocarbons given off by maturation of organic matter due to heat stored in the undercompacted zone are trapped, drilling through the source rock is accompanied by an increase in the background gas. On the other hand, selective retention of heavy hydrocarbons as a result of the migration of light components through the transition zone leads to an anomaly in gas composition.

In a zone of normal compaction, there is generally less propane (C_3H_8) than ethane (C_2H_6). When drilling into or even towards a transition zone, this relationship is often seen to reverse. In other words the C_2/C_3 ratio falls below 1. By systematically calculating this ratio, it is sometimes possible to identify the top of the undercompacted zone. Its use is strictly qualitative. The appearance of C_{3+} components can in itself be a strong indicator.

There are few published results on this subject. Basin scale statistical analyses should allow the definition of variations encountered while penetrating overpressured zones.

Gas composition analysis is affected by measuring instruments and mud characteristics. The preferential evaporation at the surface of lighter components and, inversely, the retention of the heavier components in the mud can falsify ratio evaluation. Measurements using vacuum evaporation techniques keep these disadvantages to a minimum.

Comparisons should only be made between gas shows from argillaceous layers, since reservoir gas can introduce errors, especially after recycling.

3.4.2.2. Mud Density

With modern methods for measuring density, particularly gamma ray density measurement, mud weights in and out can be monitored continuously and accurately.

A decrease in mud weight out (for a constant mud weight in) may be due to several reasons :

— expansion of the gas released by drilling of the formation as it reaches the surface,
— a kick of hydrocarbons or water (spontaneously or as the result of swabbing),
— gas diffusion (if ΔP is negative),
— a bubble of air (after tripping or connection).

The sometimes abrupt decreases in mud weight observed at the surface do not necessarily reflect a decrease throughout the annulus. Gas released at depth dissolves in the mud and only expands as free gas close to the surface. The greater the gas saturation and depth of release and the lower the mud weight, the greater will be the depth at which expansion takes place.

Figure 82 illustrates the relationship between the depth at which gas expansion occurs and mud weight.

A small gas kick at the bottom of the hole causes an expansion in mud volume at the surface which is not commensurate with the initial kick volume. For example 1 litre of gas

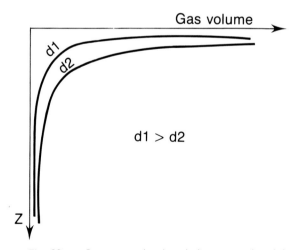

Fig. 82. — Gas expansion in relation to mud weight.

released at a depth of 4 000 m with a mud weight of 1.50 yields a volume of the order of 600 litres at the surface. This sudden expansion close to the surface is reflected by a rapid fall in mud weight, an increase in return flow, a temporary rise in pit levels and of course an increase in the gas indicated by the detectors.

Figure 83 shows the situation where a very small gas kick of 140 litres, due to drilling through a fissure, reached the surface after a lag time of 70 minutes as a sudden increase in gas together with a fall of 65 points in mud weight out and an increase of 880 litres/min in return flow.

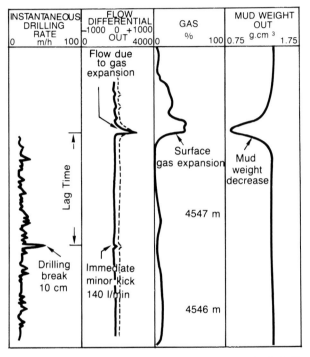

Fig. 83. — Record of gas expansion following a slight bottomhole kick (South West France).

Most reductions in mud weight are due to gas released while drilling.

The *volume of gas released at the bottom of the hole* while drilling can be calculated using the following formula :

$$V_G = (1,27\ D)^2 \cdot \pi \cdot \frac{R}{600} \cdot \phi \cdot S_g$$

where V_G = volume of gas released into the mud per minute ($l \cdot min^{-1}$)
  D = hole diameter (inches)
  R = drilling rate (m/h)

ϕ = formation porosity (decimal number)
S_g = gas saturation of the formation (decimal number)

Example : let $D = 12\ 1/4''$, $R = 30\ m/h$, $\phi = 0.3$, $S_g = 0.7$

$$V_G = (1.27 \times 12.25)^2 \times 3.14 \times \frac{30}{600} \times 0.3 \times 0.7$$

$$V_G = 7.98\ l \cdot min^{-1}$$

An estimate of the approximate *volume at the surface* can be calculated as follows;

$$V_{GS} = \frac{V_G \cdot P}{1.02}$$

where : V_{GS} = the volume of gas at the surface ($l \cdot min^{-1}$)
       P = the hydrostatic pressure of the mud ($kg \cdot cm^{-2}$)
   1.02 = atmospheric pressure ($kg \cdot cm^{-2}$)

Applying this to the foregoing example if the depth is 3 500 m and mud weight is 1.30 :

$$P = \frac{3\ 500 \cdot 1.30}{10} = 455\ kg \cdot cm^{-2}$$

$$V_{GS} = \frac{7.98 \cdot 455}{1.02} = 3\ 560\ l \cdot min^{-1}$$

The *fall in mud density* as a function of flow is calculated by the formula :

$$d_r = d_o \cdot \frac{Q}{Q + V_{GS}}$$

where : d_r = density of the gas-cut mud ($g \cdot cm^{-3}$)
       d_o = nominal mud weight ($g \cdot cm^{-3}$)
       Q = mud flow ($l \cdot min^{-1}$)

If $Q = 2\ 400\ l \cdot min^{-1}$:

$$d_r = 1.30\ \frac{2\ 400}{2\ 400 + 3\ 560}$$

$d_r = 0.52$.

In order to simplify calculations, no account has been taken here of either temperature or the compressibility coefficient, the effects of which can be ignored except when dealing with a deep well or steep geothermal gradient.

The calculations in the above example indicate major decreases in density, but in most cases these have only a minimal effect on equivalent bottomhole density. This can be calculated as follows :

Equivalent bottomhole density :

$$d_{eqv} = \frac{(P - \Delta P) \cdot 10}{Z} ;\ \Delta P = \frac{d_o - d_r}{d_r}\ ln\ (P + 1.02)$$

ie., in the above example :

$$\Delta P = \frac{1.30 - 0.52}{0.52} \ln (455 + 1.02) = 9.19 \text{ kg} \cdot \text{cm}^{-2}$$

$$d_{eqv} = \frac{(455 - 9.19)}{3\,500} \cdot 10$$

$$d_{eqv} = 1.27$$

This formula shows that the fall in equivalent bottomhole density will be greater, the shallower the depth at which the kick occurs. This is illustrated by Figure 84, which gives the relationship between equivalent bottomhole density, depth and density decrease at the surface for a mud weight of 1.50.

Where a thick gas-bearing reservoir is being drilled continuously, the mud column gas content will increase due to drilling and recycling of gas resulting in the progressive reduction in equivalent bottomhole density as the onset of expansion occurs at progressively deeper levels. Increasing the mud weight will not overcome this situation. It requires drilling in stages with intermediate circulation to degas the mud (mud conditioning).

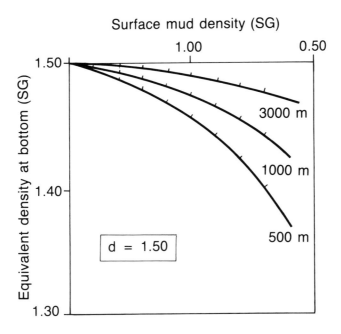

Fig. 84. — Equivalent bottomhole density as a function of density drop and depth (for d = 1.50).

Where undercompacted zones with a negative ΔP are being drilled, continuous diffusion of gas into the mud causes a progressive fall in density as the hole deepens.

Decreases in density are most frequently caused by isolated slugs of gas-cut mud whose effects need neither an increase in mud weight nor a halt to drilling and at most a short circulation.

3.4.2.3. Mud temperature

Measurements of mud temperature can be used to detect undercompacted zones and even, under ideal conditions (see below), to anticipate their approach. This is because temperature gradients observed in undercompacted series are, in general, abnormally high compared with overlying normally pressured sequences.

□ *Geothermal concepts*

The geothermal gradient, or the rate at which formation temperature increases with depth, is calculated as follows :

$$G_T = 100 \, \frac{T_2 - T_1}{Z_2 - Z_1}$$

where : G_T = geothermal gradient (°C/100 m)
T_1 = temperature (°C) at a depth Z_1 (m)
T_2 = temperature (°C) at a depth Z_2 (m)

Average geothermal gradients have been established on a regional basis and vary between 1.8 and 4.5°C/100 m in sedimentary basins (North Sea : 4°C/100 m, South West France : 2.9°C/100 m).

The geothermal flux, which represents heat flow, is determined by the following equation :

$$Q_T = \lambda \frac{\Delta T}{\Delta Z}$$

where : Q_T = geothermal flux (m W $\cdot$ m^{-2})
λ = thermal conductivity of the given formation (W $\cdot$ m^{-1} $\cdot$ °C^{-1})
$\dfrac{\Delta T}{\Delta Z}$ = geothermal gradient (°C $\cdot$ 10^{-3}m)

The geothermal gradient is not constant throughout a well, but varies according to the thermal conductivity of the various components in the sedimentary column, for example :

— Pure quartz is the mineral with the highest conductivity.

— The presence of clay minerals, especially kaolinite, considerably reduces the conductivity of the matrix.

Matrix conductivity can actually be calculated from the proportions of the different constituents :

$$\lambda_{\text{matrix}} = \pi \, \lambda_i{}^{\alpha i}$$

α_i being the proportion of constituent i in the matrix, and λ_i its thermal conductivity.

The conductivity of a clay or shale can vary by a factor of 2 depending on the nature of the constituent clay minerals. The presence of organic matter also tends to reduce thermal conductivity. Quartz on the other hand, even in microcrystalline form, greatly increases the conductivity of a clay.

On average, sediments have lower conductivity than the sialic basement. Assuming a vertical uniform flux, variations in gradient are associated with variations in thermal conductivity. The deposition of a thick sedimentary layer will act as an insulating blanket and reduce heat exchange between the basement and the surface.

It should be noted that porosity considerably decreases conductivity because of the very low conductivity of water. The nature of the fluid filling the pores (water, oil, gas) also plays a part, as gas is even less conductive. Porosity acts as a brake on heat transmission. This is particularly true in the case of high sedimentation rates which do not allow the sediments to reach thermal equilibrium with the underlying basement (LUCASEAU & LE DOUARAN, 1985).

The empirical relationship of WOODSIDE & MESSMER (1961) seems to apply very well to the range of porosities lying between 10 and 50 % :

$$\lambda_{sediment} = \lambda_{water}^{\phi} + \lambda_{matrix}^{(1 - \phi)}$$

where ϕ is the porosity.

Typical values for the thermal conductivities of the main components of sediments are :

λquartz : 7.7 Wm^{-1}°C^{-1} λchlorites : 5.0 Wm^{-1}°C^{-1} (depending on their chemistry)
λlimestone : 3.7 Wm^{-1}°C^{-1}
λdolomite : 4.5 Wm^{-1}°C^{-1} λwater : 0.65 Wm^{-1}°C^{-1}
λanhydrite : 6.0 Wm^{-1}°C^{-1}
λkaolinite : 2.1 Wm^{-1}°C^{-1}

As a result of their high porosity and hence water content, undercompacted clays behave like insulating bodies. LEWIS & ROSE (1970) have demonstrated the effect of an insulating body on heat flow (Figure 85).

The distribution of the isotherms shows a reduction in gradient on approaching the insulating body and an increase within it (Figure 86).

The fall in the rate of temperature increase on approaching an insulating body may act as a warning of the presence of undercompacted clays. An increase in temperature gradient, however, is a feature common to undercompacted zones and other insulating formations, such as very porous reservoirs, or thick coals.

□ *Measuring mud temperature*

Several methods are used to measure mud temperature, their objective being to obtain a best approximation to the formation temperature. For the sake of uniformity all the methods, whether used during or after drilling, have been grouped together below :

Surface

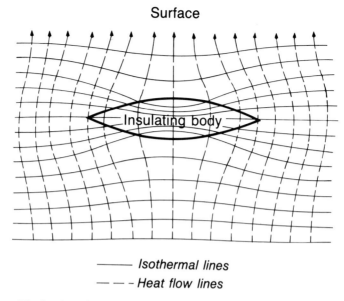

—— Isothermal lines
— — - Heat flow lines

Fig. 85. — Distribution of heat flow lines and isotherms around an insulating body
(modified from Lewis & Rose, 1970).

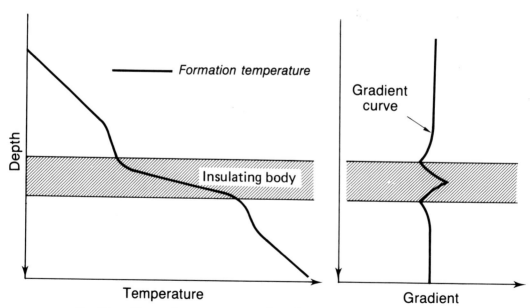

Fig. 86. — Change in temperature and gradient across an insulating body
(modified from Lewis & Rose, 1970).

● *Surface measurements :*

Surface measurements are continuously recorded by sensors (pyrometers) placed in the mud-return line for the temperature out, and in the suction pit for the temperature in. Pyrometers work on the principle of measuring changes in the resistance of a platinum wire which forms part of a Wheatstone bridge.

As the temperature at the surface is affected by many climatic and human factors (additions of mud, etc.), the differential temperature can be used to eliminate fluctuations in the return line temperature caused by fluctuations of the inflow temperature. The rise in temperature for a given mud sample can be evaluated when the lagtime is known (Fig.87).

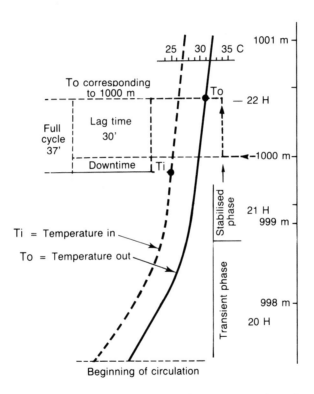

Fig. 87. — Method for correlating mud temperatures in and out.

Calculation of the differential temperature is very much easier when a computerised mud logging unit is available.

Because of the rapid circulation of the drilling mud and the effects of forced convection, the measured temperature profile differs from the actual geothermal profile. The mud in the upper part of the hole is warmer than the formation, and in the lower part it is cooler (Fig.88).

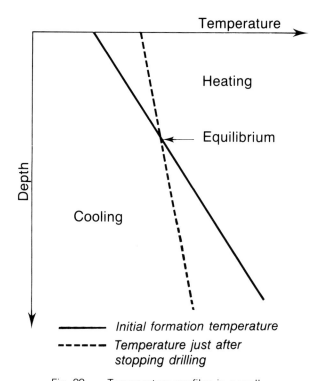

Fig. 88. — Temperature profiles in a well.

The thermal profile established in a well while drilling depends essentially on the following factors (CORRE *et al.*, 1984) :

— inflow temperature, which depends on the amount of *cooling at the surface,* which is generally a few degrees (between 1 and 5°C),

— the *rate of inflow,* which acts in two ways :
   — it controls the speed at which the mud and the calories it contains return up the annulus,
   — together with the *pump pressure,* it controls the hydraulic energy fed into the system, which also heats the mud ;

— the *thermophysical properties of the mud,*

— the *bottomhole temperature,* itself a function of true formation temperature and heating effects at the bit face.

The thermal profile is not very sensitive to local variations in geothermal gradient or drilling rate.

The mechanical energy expended while drilling plays a part, especially in heating the bit, but its effect on mud temperature is virtually negligible in comparison with hydraulic energy.

To give a numerical example, take a 17 1/2″ diameter hole with a mud flow Q of 3 700 l · min⁻¹, a pump pressure P of 230 bar, a rotation speed of 100 rpm with a torque of 500 m · kg⁻¹. This produces hydraulic energy of :

$$E_h = P \cdot Q = 1\ 418\ KW$$

and mechanical energy of :

$$E_m = 2\pi \cdot C \cdot \theta = 60\ KW$$

This can explain why shallow wells drilled with very large mud flow rates frequently yield abnormally high temperatures, caused by significant heating as the mud passes the jets in the bit, where the major pressure loss occurs.

All these factors associated with drilling predominate over the formation temperature itself.

The table below (fig.89) refers to a 5 000 m hole. It compares the temperatures measured at the surface with the maximum recorded in the course of wireline logging. Return mud-line temperatures when drilling at 900 m and 5 048 m have comparable values, while the wireline maxima differ by 130°C. In this example, flow effects can clearly be seen in the upper part of the well. Over the well as a whole the impact of mud-flow variations is more marked than that of heat exchange at the walls. A reduction in the flow from 1 700 to 650 l · min⁻¹ produces a fall of 14°C in return line temperature, while maximum temperature increases by 24°C over the same interval.

Depths (m)	Hole diameter	Mud flow rate (l · min⁻¹)	Pump pressure (bar)	Wireline log temperature	Mud temperature out
900	17 3/8	3700	95	35	37
2582	12 1/4	2200	165	71	56
4625	8 1/2	1700	130	126	60
4850	5 3/4	650	135	150	46
5048	5 3/4	650	135	165	40

Fig. 89 — Table summarizing temperatures observed while drilling — South West France.

M.W.D. tests in the North Sea have shown that a reduction in flow may in fact be accompanied by an appreciable fall in bottomhole temperature while drilling (Figure 90).

Despite the mechanical and hydraulic effects, it is possible *under ideal conditions* to obtain variations in mud temperature which are representative of geothermal variations.

Figure 91 illustrates an onshore case in Nigeria where undercompaction is revealed by a change in mud temperature out.

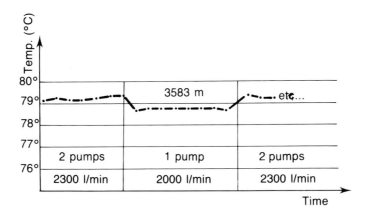

Fig. 90. — Effect of mud flow rate on bottomhole temperature (M.W.D.).

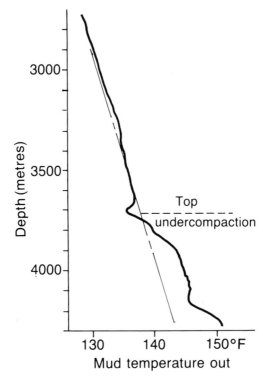

Fig. 91. — Detecting an undercompacted zone from the mud temperature out (Nigeria).

Such cases, where mud temperature out has actually been used to detect under-compacted zones, are in fact fairly rare. Changes such as those observed in Figure 91 are frequently masked for various reasons :

— *offshore drilling* : the marine riser assists heat exchange between the mud and the surrounding sea. The amount of cooling depends on the length and diameter of the riser, mud flow, string rotating speed and the sea/mud temperature contrast. In a warm sea with no strong currents the contrast effect is less than in a cold sea with variable currents.

Recently developed techniques can be used to measure the temperature at the base of the riser and so avert the effects of cooling.

— *drilling and circulation halts* : these cause cooling of the mud in the circulating pits and in the upper part of the well.

The length of the halt determines the amount of cooling and the time taken for the temperature to restabilise when drilling is resumed. To this extent longer bit runs enable mud temperature to be used more effectively.

Trend-to-trend plotting of mud temperatures out will remove irrelevant scatter and takes account of stabilised temperatures only (Figure 92).

— *Surface operations* : transfers of mud between active pits and reserve pits disturb the mud temperature in, due to temperature contrasts and changes in working volume. The many additions and transfers needed while drilling, particularly in large diameter holes, generally make interpretation of temperatures very difficult despite the use of differential

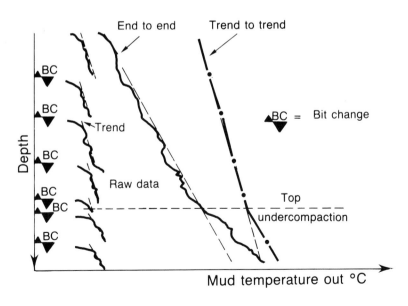

Fig. 92. — Different ways of plotting mud temperature out.

temperature. It is always advisable to note any major operations concerning the mud on the temperature diagrams.

— *Climatic changes :* in the case of an onshore well, exposure of the pits to the open air can result in significant mud temperature in variations due to the ambient conditions (sun, wind, frost, rain).

— *String rotating speed :* rotation of the string is transmitted to the mud and has an appreciable effect on thermal transfers at the borehole walls.

— *Lithology :* in order to keep lithological effects to a minimum, preference must be given as far as possible to temperatures relating to shales.

— *Fluid kick :* an influx of fluid at formation temperature will bring about an increase in mud temperature commensurate with its volume.

— *Influx or diffusion of gas :* increase in gas volume close to the surface will bring about a temperature reduction through endothermic expansion.

—¹ *Mud type :* heat exchange between the formation and the mud will depend on the conductivity of the mud. This will vary according to its nature (water, oil, polymer, etc.). Internal heating of the mud will depend on its specific heat.

— *Measurement quality :* the sheathed pyrometers generally used have a satisfactory accuracy of the order of 0.1 % if they have been calibrated correctly. Measurement quality may however be adversely affected by : the position of the rod, particularly in the case of mud temperature out ; the mud level in the return line ; turbulence and settling of cuttings around the rod. These problems can be limited if the operators supervise the equipment properly. Positioning the measuring rod further upstream, near the bell nipple, can overcome these disadvantages.

— *Plotting measurements :* poor graphical representation of temperatures can adversely affect the way they are interpreted. Some types of plot reveal noteworthy changes more clearly than others. It is better to use them all : raw mud temperature out, ΔT, end-to-end or trend-to-trend, and temperature out gradient (°C/100 m). The scales used should be linear.

The interpretation of mud temperature out should be regarded as qualitative. It may perhaps contribute to locating the top of the undercompaction and, in favourable circumstances, the approach to it. It is unlikely that it could ever yield an estimate of pore pressure.

- *Measurements While Drilling (M.W.D.)*

M.W.D. techniques can be used to measure bottomhole mud temperatures while circulation is in progress. As the mud has only had brief contact with the formation at the measure point, heating due to hydraulic effects and the work of the bit predominates.

These methods have come into use only recently, and it has not yet been possible to estimate their contribution to temperature interpretation.

- *Bottomhole measurements during wireline logging*

Whenever a wireline log is run a maximum thermometer (sometimes 2) is attached, generally to the bridle. These measurements can be used to determine the mud temperature at a given depth after circulation has been stopped for a known time. Despite time since circulation frequently in excess of 24 hours, the temperature of the mud does not reach the true formation temperature. Several logging runs at a given depth make it possible to monitor the rise in measured temperature towards formation temperature with time. A Horner plot is used to extrapolate measured temperatures. The method is based on the assumption that the mud cools the formation during drilling or circulation. This sets up a temperature gradient between the walls and the surrounding formation. When circulation is stopped, the heat exchange between the formation and the mud tends to reduce the radius of the cooled zone and thereby the gradient. By extrapolating the temperature to infinite time it is possible to deduce the true formation temperature, provided circulation was not continued too long after drilling stopped (implying excessive cooling) (Fig.93).

As a first approximation, the temperature/time relationship is as follows :

$$T = T_f - C \log \frac{t_k + \Delta t}{\Delta t}$$

where
T = measured temperature
T_f = true formation temperature
C = constant
t_k = bottom circulation time
Δt = time elapsed between stopping circulation and logging tool on bottom prior to logging.

Some authors recommend that the drilling time for the last metre should be added to t_k.

A plot of T and $\dfrac{t_k + \Delta t}{\Delta t}$ on semi-log paper is linear. Extrapolating the graph to a time factor of 1 provides an estimate of the formation temperature.

It must remembered, however, that this method is only an approximation of the actual heating curve. It has the merit of simplicity, but is valid only if the following condition is fulfilled (LUHESHI, 1983) :

$$\frac{a^2}{4 kt} << 1 \quad \text{(in practice} < 0.07)$$

with
a^2 = square of borehole radius (m²)
k = diffusivity of the system (m² · s⁻¹)
t = time for which circulation stopped (s)

In practice this condition is only fulfilled for an 8 1/2" hole. In other instances the formation temperature will be underestimated.

Although two temperature/time pairs are sufficient to draw the straight line it is preferable to have three or four. The logging geologist must check that each tool has at least one maximum thermometer appropriate to the expected temperature range. The thermometers should be reset prior to each run.

Figure 93 shows an example of extrapolation by this method.

Some authors have recommended other mathematical methods (NWACHUKWU, 1976). Their results are comparable to those obtained using the Horner plot, which has the advantage of being easier to handle.

DIAGRAPHIES DIFFEREES
WIRELINE LOGGING

X X X 1

RUN N° : 3

Extrapolation de la température
Extrapolated temperature

Compagnie de service : *Contractor :*	Date : 05.11.84

Profondeur forage : 2620 m
Driller depth

Arrêt forage : 5 november 18h 00
Stop drilling

Arrêt circulation : 5 november 21h 30
Stop circulation

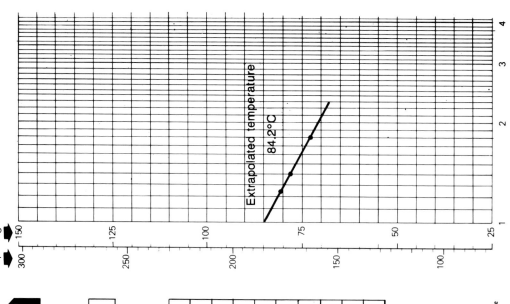

Extrapolated temperature
84.2°C

$(tk + \Delta t)/\Delta t$

Type d'outil *Tool type*			
Prof. log *Depth logger*	2615	2615	2615
D	5-21.30	5-21.30	5-21.30
tk1	20	20	20
tk2	210	210	210
t	6- 2.20	6- 7.30	6-11.50
$\Delta t = t - D$	4.50	10.00	14.20
tk = tk1 + tk2	3.50	3.50	3.50
Temp. max. enreg. *Max. rec. temp.*	73.5	78.2	80.1
$\dfrac{\Delta t + tk}{\Delta t}$	1.79	1.38	1.27

D — Date fin de circulation (J-H-Min.)
Date end of circulation (D-H-Min.)

tk1 — Temps mis pour forer le dernier mètre (Min.)
Time to drill the last meter (Min.)

tk2 — Temps de la dernière circulation (Min.)
Time of last circulation (Min.)

t — Date début remontée outil (J-H-Min.)
Date of starting pulling out (D-H-Min.)

MODE OPÉRATOIRE
OPERATING MODE

Reporter pour chaque outil $\dfrac{\Delta t + tk}{\Delta t}$ en fonct. de temp. max. La température extrapolée est obtenue par l'intersection de la droite moyenne avec l'axe des ordonnées.
Plot for each tool $\dfrac{\Delta t + tk}{\Delta t}$ related to the max. rec. temp. The extrapolated temperature is obtain by intersection of, the trend with the vertical axis.

Fig. 93. — An example of temperature extrapolation.

The use of extrapolated temperatures provides an approximation to the value of the geothermal gradient. Figure 94 is an example from the North Sea where the undercompacted zone is revealed by an increase in the geothermal gradient. In this case, the information provided by extrapolated temperatures is more meaningful than information from surface mud-temperature measurements, which are strongly affected by a reduction in flow in the riser (greater cooling).

- *Measurements while running wireline logs*

A Schlumberger tool, the AMS, which is compatible with most logging tools, provides a continuous record of temperature during its descent (the mud being less disturbed than when it is pulled up). Temperatures measured in this way in a North Sea well have revealed appreciable differences from the temperatures measured using maximum thermometers.

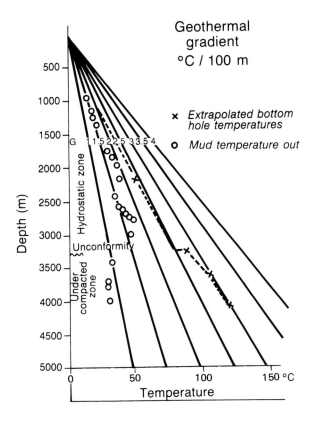

Fig. 94.— A sample comparison between extrapolated bottomhole temperature measurements and surface mud-temperature measurements (North Sea).

● *Bottomhole measurements during formation testing*

Temperature measurements of fluid produced in the course of formation testing are more representative of formation temperature. Two types of measurement may be performed :

— a maximum thermometer of the type used in wireline logging is placed in the mechanical pressure recorders (Amerada, Johnston etc),

— continuous thermometry in association with electronic pressure recorders.

Such measurements are not performed regularly and it is not common for them to be made at several depths within a given well.

The excess pressure exerted by the mud column on the borehole walls creates a zone of thermal disturbance around the well through invasion of the formation by mud filtrate. For this reason a long period of flow is required in order to obtain a temperature measurement representative of formation temperature.

In South West France the following results have been obtained for a geothermal well :

— Test no 1 at 1 278 m Flow 9 600 litres (after drilling with partial mud losses) Bottomhole temperature = 37°C

— continuous thermometry recorded 4 weeks after the production tests Bottomhole temperature (station at 1 227 m) = 46.05°C.

● *Bottomhole "Temp Plates" measurements*

The geothermal gradient between two logging runs can be measured using self-adhesive temperature indicators called Temp Plates. These change colour under the effect of temperature due to a chemical reaction. They are generally placed on directional tools (TOTCO). They can be accurate to ± 1°C provided the optimum temperature range is chosen.

As measurements are made within the string shortly after stopping circulation, the temperatures obtained are not very representative of true formation temperatures.

● *Thermometry*

A continuous profile of the change in temperature of the mud column in a well can be obtained with thermometric logging tools (HRT, etc.). They are based on the principle of measuring changes in the resistance of a cadmium wire mounted in a Wheatstone bridge. Their sensitivity is in the range 0.1 to 0.3°C.

These techniques are more generally used either in geothermal wells or in petroleum drilling to detect mud-loss zones or the top of cement behind a casing.

It should be remembered that measured values are not representative of either the formation temperatures or changes in the gradient, because mud temperatures are not stabilised in relation to the true formation temperatures (Fig.88).

CONCLUSION

Although undercompacted zones are accompanied by temperature anomalies, it is not easy to detect these using available methods for measuring mud temperature. These methods should in theory be capable of detecting them while drilling, or even before they are reached, but they depend on a number of variables which frequently mask changes in geothermal gradient.

The low success rate for these cheap and easily applied methods may be improved by greater care in the installation and maintenance of measuring equipment and by more complete and careful interpretation of the data acquired.

Bottomhole temperature measurements during logging, which have the disadvantage of being performed subsequently and in isolation, nevertheless provide a better estimate of true formation temperature. However, the quality of the measurement depends directly on the time elapsed since drilling ceased.

3.4.2.4. Mud resistivity

In a normally compacted series interstitial water salinity increases gradually with depth. OVERTON & TIMKO (1969) have demonstrated the role of ionic filtration by clays in the course of water expulsion during sediment compaction. The retention of ions within the clay gives rise to an increase in pore water salinity, whereas the salinity of expelled water is often taken to be zero. Measurements made in the Gulf Coast have revealed a close relationship between the progressive changes in porosity and pore water salinity with increasing depth (Fig.95).

MAGARA (1978), revising the results of OVERTON & TIMKO, demonstrated that because ionic filtration is not perfect, the salinity of the expelled water was not absolutely zero. In fact if it is assumed that clays have perfect ionic efficiency, the ion concentration per unit volume of clay increases with compaction (a volume at the surface being considerably reduced at depth). Finding that the volume of dissolved salts per unit volume of clay was constant regardless of the depth, MAGARA deduced that part of the salt is lost during compaction. Figure 96 shows, on the basis of the same data, the salinity graphs which would result if either completely pure water (A) or water having a salinity equal to that of sea water (B) were expelled. As the graph for the actual measurements lies between these two assumptions, its slope depends on the ionic filtration efficiency of the clay and therefore on its composition. A montmorillonite, for example, is a more effective filter than an illite because of its high ion exchange capacity.

Monitoring changes in the salinity of formation water should therefore theoretically make it possible to detect undercompacted zones. Equipment can be used to measure mud resistivity/conductivity continuously while drilling. Fitting resistivity meters on the mud inflow and return lines can be used to evaluate the differential resistivity. This can be translated approximately into terms of chloride content.

The detection of changes in salinity while drilling requires a significant contrast in resistivity between mud and formation water. The release of formation water by drilling alone is insufficient in comparison to the volume of circulating mud to give rise to measurable

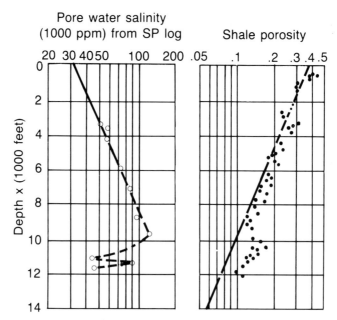

Fig. 95. — Relationship between depth and clay porosity — salinity (Magara, 1978, after Overton & Timko, 1969, reprinted by permission of the American Association of Petroleum Geologists).

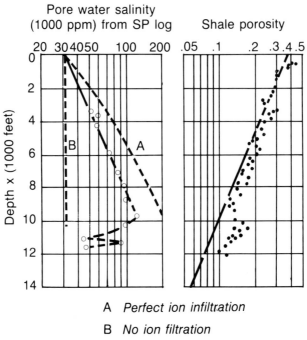

A *Perfect ion infiltration*

B *No ion filtration*

Fig. 96. — The effect of ion filtration efficiency on salinity (Magara, 1978, after Overton & Timko, reprinted by permission of the American Association of Petroleum Geologists).

changes in resistivity. Only kicks or continuous diffusion of formation water into the well, due to negative ΔP, will show up as significant changes in resistivity. This effect will be more appreciable the less saline the mud.

3.4.3. *CUTTINGS ANALYSIS METHODS*

3.4.3.1. Lithology

The expected lithological sequence may provide an overall indication of the possible existence of abnormal pressure. The presence of seals, drains or thick clay sequences is a determining factor in this analysis.

Depending on the regional geological context, the identification of clay, evaporite or compact limestone horizons provides geologists with a warning of the likely presence of abnormally pressured zones. In the case of an exploration well, entry into a porous zone is generally marked by a drilling break after passing through the cap rock, and should be monitored attentively by observation of the well and complete geological circulation (bottoms up).

As far as undercompaction is concerned, the risk of abnormal pressure depends in essence on the thickness of the clays. In a thick clay sequence undercompaction is frequently encountered shortly after penetrating the clay. If drains are present within the transition zone the increase in pressure tends to be more gradual. Nevertheless cases have been observed where thin clay bands (10 to 20 m) act as barriers between thick drains (in excess of 100 m) marked by very different pressure gradients (Nigeria).

Clastic sediments (sand) or bioclastic sediments (chalk, grainstone) may be undercompacted in these environments. The progress of their compaction is more difficult to identify than in the case of clays, given the variations in grain size, cementation, etc.

Analysis of the composition of the clays, and more particularly of their silt and carbonate contents, contributes to a better interpretation of undercompaction detection parameters. For example the drilling of a silty clay is slower if the silt is dispersed within the clay matrix than if it is present in the form of thin laminations. The influence on "d" exponent etc. must be taken into consideration. Changes in texture may themselves constitute a detection factor. The reappearance of plastic clay (increase in smectites) after passing through a series of indurated clays may suggest undercompaction.

A mineralogical change in the clay itself, normally undetectable with a microscope except perhaps indirectly from the texture, may be envisaged if laboratory techniques such as X-ray diffraction are adapted for use at the wellsite.

3.4.3.2. Shale density

Measurement of clay and shale density is one of the oldest methods of detecting abnormally pressured zones. It is based on the principle that shale density in an undercompacted zone increases less rapidly and may even fall in comparison with the density of normally compacted overlying clays and shales.

The effectiveness of this method, which is still in use, depends on the the cuttings selected being representative.

The validity of the apparent densities obtained depends on the following factors :

● *consolidation* of the clay : in Cameroon, for example, where drilling seldom goes below a depth of 2 000 m, the clays are not sufficiently consolidated to allow their density to be measured under current wellsite conditions.

● *clay composition* : the presence of accessory minerals (pyrite) and changes in the concentration of silt and carbonates have a great influence on the measured density. For example, given a shale with a density of 2.28 and a water-filled porosity of 15 %. With a 10 % pyrite content it would have a density of 2.52.

● *depth lagging* : the cuttings selected must be representative of their drilled depth. Rapid drilling with caving can make selection difficult. In a uniform argillaceous series, selecting cavings can falsify measurements. The problem may be reduced by more frequent measurements, which will provide a statistically better approximation to the density.

● *mud type* : the use of reactive muds, which is most frequently the case with water-based muds, has an adverse effect on measurement quality. The cuttings adsorb water while rising up the well-bore, particularly when smectite content and porosity are high. In these circumstances measured densities are lower than actual densities.

Non-reactive mud (oil-based muds) do not have this disadvantage. Non-dispersant muds, which are inhibited against clay swelling and dispersion by one or more electrolytes, have been prepared to solve the problems inherent in drilling water-sensitive clays, whereas dispersant muds give rise to swelling, dispersion of the cuttings and erosion of the hole walls. Most non-dispersant muds contain potassium salts which have the property of reducing osmotic hydration of the clays and hardening them without completely dehydrating them.

● *methods of measurement* : different methods exist — the most commonly used are the following :

— heavy liquids :

Kits of liquids of different density are available commercially or can be made up on site using two miscible liquids of different density and a Mohr balance. A set of densities from 2.20 to 2.70 g · cm^{-3} in stages of 0.05 g · cm^{-3} will cover the entire range of shale densities. The method is based on Archimedes principle. Each cutting is immersed successively in liquids of increasing density until it no longer sinks.

The speed with which the cutting sinks (in the densest liquid in which it does sink) is observed in order to obtain an accuracy greater than 0.05 g · cm^{-3}.

However, in addition to the disadvantage of insufficient accuracy, cuttings must be transfered from one liquid to another with care in order to avoid any change in the density of the liquids through mixing.

— variable density column :

A variable density column can be prepared by partially mixing miscible liquids of known densities. The density distribution is checked using beads of calibrated density which can be used to prepare a graph of density against column height. To make the technique easier to use it is preferable to have a linear relationship, which can be obtained by careful mixing of the different liquids.

The most commonly used liquids are bromoform (d = 2.89) and carbon tetrachloride (d = 1.59), or the somewhat less toxic trichloroethylene (d = 1.47).

Each cutting, taken separately, is immersed in the column after having been dried on absorbent paper, then all that is necessary is to read off the height at which the sample comes to a halt and check this value on the calibration curve in order to obtain the shale density.

As long as the column is properly calibrated this method has the advantage over the previous method in that it is more accurate and faster. It is the most widely used method.

— mercury pump :

The total volume of a sample of cuttings of known weight is measured using a high pressure mercury pump (the so-called Kobe method).

After its weight has been measured using an accurate balance the sample placed in the mercury pump is pressurised, a procedure which provides a measure of its volume without contact between the cuttings and the mercury.

The advantages of this method are the accuracy inherent in the equipment itself, and statistical representativeness due to the amount of cuttings used (about 25 g).

On the other hand the possible presence of a large volume of cavings, which are sometimes difficult to separate, constitutes a source of error. Care must be taken when manipulating the apparatus to avoid contact with the mercury, which is poisonous.

This method is relatively slow in comparison with the foregoing methods : sample preparation may be laborious where the lithology is complex.

— pycnometer :

This method involves placing the sample in a container of known volume and measuring the change in weight of the container after fluid has been displaced by the sample.

This method is easy to apply at the wellsite using a mud balance. Cuttings are placed in the empty pan of the balance until it reads 1 g $\cdot$ cm^{-3} (with stopper in place). The pan is then filled with water and the new density is noted (d$_2$).

The apparent density is obtained from the following formula :

$$\rho_b = \frac{1}{2 - d_2}$$

where ρ_b = apparent density of the clay
d_2 = density reading after the addition of water.

Although the mud balance is simple and easy to use, it does not give very accurate results. But in the absence of other methods it may be a useful stop-gap.

— "Microsol" (Geoservices) :

The principle behind this method involves comparing the weight of the cuttings in air and in water. Shale density is obtained by the formula :

$$\rho_b = \frac{L_1}{L_1 - L_2}$$

where L_1 = weight in air
L_2 = weight in water

Three or four measurements are made every 5 m. The arithmetic mean is taken to be the density.

Use of the Microsol is difficult, particularly offshore where it should not be used because of friction problems arising between the float and the walls of the test tube.

All of these methods require special treatment of the cuttings. Washing in every case and (in the case of the dense liquids, column and mercury pump methods) drying of the surface *without heating* (in order to avoid dehydrating the clays).

Where the dense liquids or column methods are used the cuttings should be selected to remove fissured fragments which can retain air. Observation of the background gas should draw attention to low densities which might result from the presence of gas in the shales.

Figure 97 illustrates an instance of undercompacted clays from offshore Nigeria. The information obtained from the density plot is excellent. When it was first used, attempts were

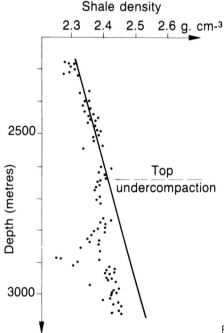

Fig. 97. — Shale density (Nigeria).

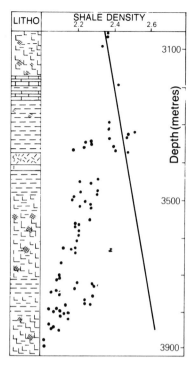

Fig. 98. — Shale density and lithology — Triassic evaporite/shale section — South West France.

even made to use this method for evaluating pressures quantitatively (Chapter 4), but it gave mediocre results.

In the example in Figure 98, from the Aquitaine Basin, thin bands of highly under-compacted clays are isolated by alternating impermeable layers of salt, anhydrite and carbonates.

Shale densities may prove very useful regionally, especially if the anomaly has been calibrated by a check with wireline density logs.

It should be noted that the various methods of measuring the apparent density of shales may be applied to other lithologies to help determine the overburden gradient while drilling (see Chapter 4.2).

3.4.3.3. Shale factor

The importance of the cation exchange capacity (CEC) or water adsorption capacity has been mentioned in section 2.3 (clay diagenesis).

The principle is as follows. A clay mineral such as kaolinite, illite, smectite, etc., reacts with the surrounding hydrated environment in relation to its accessible specific surface area on which cation exchange sites are located.

This ability to exchange cations is a specific characteristic of clays in general. The cation exchange capacity, commonly expressed in meq (milliequivalents) per 100 g of clay, has a specific mean value for every clay type.

For example, the following figures are conventionally used :

— Kaolinite : 5 meq/100 g

— Illite : 10-40 meq/100 g

— Smectite : 100 meq/100 g

A gradual reduction in CEC with depth, reflecting the transformation of smectites into illite, is generally observed (Fig.99).

Methylene blue is a polyelectrolyte which fixes itself on the exchange sites of a clay. If the amount of fixed methylene blue is known the number of active sites on a clay can be calculated.

An approximation to CEC may be obtained at the wellsite using a method familiar to mud engineers for many years in order to estimate bentonite (smectite) content in the mud by methylene blue titration). This procedure determines the shale factor. Although they cannot be used to calculate the proportions of each of the clay minerals, the measured values can be used to represent a change in relation to depth.

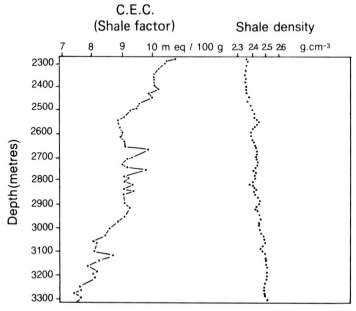

Fig. 99 — Change in shale factor and shale density with depth (example from Nigeria).

The method to be used on site is as follows :

— when the cuttings have been dried, sort them and select those which are representative of the new formation,

— grind the cuttings finely using a mortar,

— weigh 0.5 g of this powder, add distilled water and a few drops of sulphuric acid (5N) in a dish,

— heat to boiling, stirring the suspension continuously,

— add the methylene blue solution drop by drop taking one drop of the mixture at regular intervals and placing it on a filter paper.

— if the colour remains in the centre of the spot obtained the equilibrium point has not yet been reached.

— if the colour follows the enlargement of the spot and forms an aureole around its periphery, note the volume of methylene blue used.

— the shale factor is calculated as follows :

$$\text{shale factor} = \frac{100}{W} \times V \times N$$

where W = weight of powdered rock (grams)
      V = volume of methylene blue used (ml)
      N = methylene blue concentration.

The expression "shale factor" must be used in this case in order to distinguish these results from the true CEC.

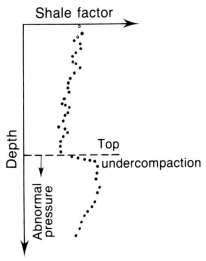

Fig. 100. — Diagrammatic change in the shale factor in an undercompacted zone associated with a smectite increase.

Impurities, the procedure itself and measurement errors permit a wide margin of inaccuracy. The fact that the methylene blue molecule is too large to be adsorbed onto interlayer sites makes the results consistently lower than the true values.

Figure 100 provides a diagrammatic illustration of the behaviour of the shale factor in a case of undercompaction resulting from the overburden effect, associated with an increase in the smectite content within the undercompacted shales.

Undercompacted clays which have been unable to dehydrate often have an unusually high proportion of smectite and an abnormally high shale factor. However, the initial proportions of the clay minerals in the deposit can mask changes in shale factor related to pressure, or at the other extreme, exaggerate them and give false alarms.

Shale factor is not a reliable technique for detecting abnormal pressures, and cannot on its own lead to the conclusion that they are present. It may however provide confirmation and assist interpretation. The results may also contribute to the recognition of lithological markers.

3.4.3.4. Shape, size and abundance of cuttings

Large cuttings are generally regarded by wellsite geologists as being cavings. But we have seen that in under-balanced drilling (negative ΔP) large cuttings may also be produced at the bit face and be confused with cavings. A concomitant disappearance or sharp reduction in very fine cuttings can generally be used to decide the matter.

More generally the presence of a large volume of cavings implies instability of the borehole walls. In other words an increase in the volume and the size of the cuttings is an indication of thermal or mechanical imbalance when drilling.

The problem is mainly associated with argillaceous rocks, although all other formations may also be affected, from carbonates to volcanic rocks, provided that they are located at sufficient depth.

Cavings are the result of breakdown of the walls due to excess stress, but current research in this subject shows that there are many modes of fracture, which makes interpretation of the shape of cavings difficult at the present time. It can nevertheless be held that stress in the walls can reach an upper compression limit or a lower tension limit.

High formation pressures contribute to destabilisation of the borehole walls in two essential ways. On the one hand, they reduce the strength of the rock, while on the other they can cause circular concentric tension fractures in low permeability formations such as shale. This happens when there is strong hydraulic pressure imbalance between the formation and the mud in the well which cannot easily be relieved due to the low permeability. The radial pressure gradient which is then set up between the formation and the mud can exceed the tensile strength of the matrix, causing these fractures. Improved bit performance in zones of high formation pressure is also partly due to the same process.

The very creation of a borehole (or tunnel) creates excess stress at the walls. The mechanical effects procured by the drill string and the mud will obviously accelerate the destabilization process already begun. Cavings produced by this process are larger than

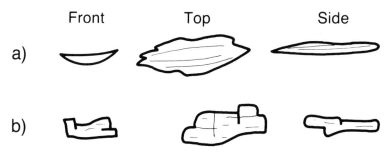

Front Top Side

a)

b)

Fig. 101. — Typical caving shapes. a) flaky b) blocky

those simply produced by erosion, which can often be confused with cuttings from the bit face.

The cavings observed on the shale shaker have two essential shapes (see Figure 101 a/b). The first is a flattened, elongated flake, frequently confused at first sight with the cleavage of a laminated shale. When examined closely it reveals a concave cross-section, which rules out the latter possibility. The second shape is more blocky, often with microfissures.

Laboratory tests have demonstrated that the fracture mechanism due to excessive compression can produce both types of cavings at the same time or in succession.

Plate-shaped cavings are therefore not a definite indication of overpressure, since they can also be created by stress effects in normally compacted rocks.

Recent research combining the wellsite study of cuttings shapes to create a "cavings log" as part of a geomechanical surveillance program, together with laboratory studies, may allow a better understanding of caving mechanisms and their true relevance to detecting overpressured zones.

3.4.3.5. Cuttings gas

For many years some mud logging companies have offered a cuttings gas or "occluded gas" service. This involves the wellsite measurement of the percentage of gas released from a given volume of cuttings by breaking them up in a kitchen-type blender. Restricted to catalytic filament detection and without the assistance of chromatography, this has rather fallen out of use even though the method can provide helpful information.

Cuttings incorporate a microporous system containing formation fluid which is not polluted by the mud because of capillarity and adsorption forces. The non-polluted volume depends on the permeability of the rock. Shales retain a large proportion of their fluid content right up to the surface.

We have already seen that many factors affect the measured amounts of gas contained in the mud (3.4.2.1), so any method of cuttings gas analysis has the great advantage of being able to reveal the volume and composition of the in situ fluids.

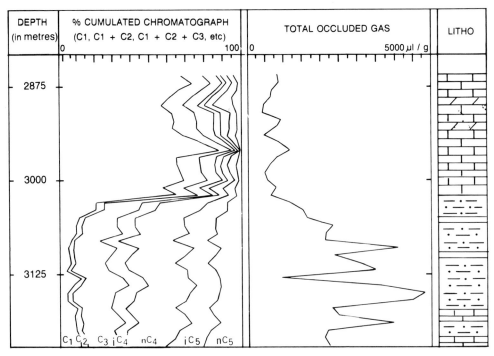

DEPTH (in metres)	% CUMULATED CHROMATOGRAPH (C1, C1 + C2, C1 + C2 + C3, etc)	TOTAL OCCLUDED GAS	LITHO

Fig. 102. — Example of a cuttings gas recording (South West France).

Techniques previously used in the laboratory have been introduced at the wellsite by the combined use of flame ionisation chromatography (FID), vacuum blending and more accurate measurement of the cuttings volume. As a result it is now possible to prepare logs similar to the one shown in Figure 102. In this example a change in lithology is revealed by the measured levels of cuttings gas. At 3 025 m and below, these show that the shale is a source-rock. Confirmation comes from chromatographic analysis, which reveals a considerable reduction in light components (methane, ethane) with consequent enrichment in the heavy components (butane and pentane).

The frequently noted increase in gas content in undercompacted clays will be better detected by this method. Similarly changes in the composition of gas indicators which frequently occur in transition zones may provide a means for the detection of abnormal pressure.

This potentially useful method, adapted to the wellsite by Elf Aquitaine and Geoservices, provides an original approach to the interpretation of gas shows while overcoming the uncertainties inherent in measuring drilling mud gas content.

158

3.4.3.6. X-ray diffraction

The recent development of portable X-ray diffractometers which can be used to determine percentages of mineralogical components has made their use possible at the wellsite, where it becomes a potential detection tool.

The differences in the interreticular spaces between montmorillonite (1.4 nM) and illite (1.0 nM) make on-site mineralogical analysis of clays a possibility. A semi-quantitative method for determining percentages of clay minerals based on the heights of recorded peaks should make it possible to determine the smectite/illite ratio. As we have seen, this ratio is often useful for the interpretation of undercompacted zones (Chapter 2).

Whether this method can replace the shale factor as an indicator of clay mineralogy remains to be seen.

3.4.3.7. Oil show analyser

The oil show analyser developed by the Institut Français du Pétrole (ESPITALIÉ et al., 1984) is a device which provides a geochemical log of penetrated strata (quantities of oil and gas, residual petroleum potential of the kerogens, organic carbon contents, etc.). This log indicates source-rocks, their state of maturity (oil zone and gas zone) and the type of organic matter which they contain, as well as identifying horizons which are impregnated to a varying extent with hydrocarbons.

The device, which is derived from the Rock Eval pyrolysis technique, is particularly adapted to the analysis of cuttings and cores at the wellsite itself. Each analysis lasts about 20 minutes and includes the following stages :

1) determination of the following quantities in a pyrolysis furnace under inert atmosphere :

— free hydrocarbons present in the sample, namely : gas (S_0 peak) and oil (S_1 peak), Figure 103a.

— hydrocarbon compounds (S_2 peak) originating at a programmed temperature between 300 and 600°C from thermal cracking of the organic matter not yet transformed into petroleum (kerogen). The temperature at which maximum release of these hydrocarbon compounds occurs, Tm, is also recorded. Quantities are expressed as mg HC/g of rock (it should be noted that asphaltenes, which are not part of the kerogen, also give an S_2 peak).

2) determination of the total organic carbon content (as a %). The amount of residual organic carbon (remaining after pyrolysis) is found by oxidation of the pyrolysis residue in a second furnace (the CO_2 produced forms peak S_4 in Figure 103a). The amount of pyrolised organic carbon is found from the hydrocarbon quantities in peaks S_0, S_1 and S_2. These values are added together. Figure 103b shows the fractions of the total organic matter which this method analyses.

These parameters are calculated by a microprocessor which can also be used to provide :

— the value of the hydrogen index (HI), which is nothing more than the amount of S_2 hydrocarbons in relation to the total organic carbon content. It has been demonstra-

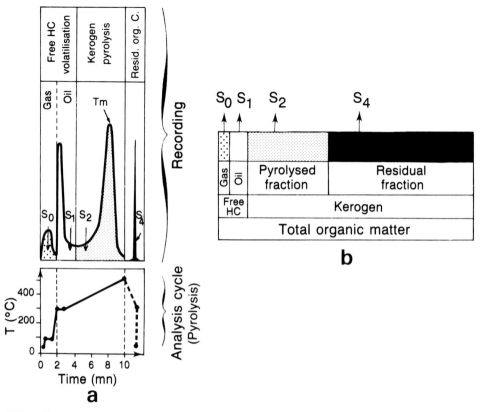

Fig. 103. — Example of an oil show analyser record (a), fractions of the total organic matter analysed (b) (Espitalié *et al.* 1984).

ted that there is a good correlation between this hydrogen index and the atomic H/C ratio obtained by elemental analysis of the corresponding kerogens.

— values of the gas production index (gas PI $= S_0/(S_0 + S_1 + S_2)$ and oil production index (oil PI $= S_1/(S_0 + S_1 + S_2)$. In the absence of migration these indices represent the degree of transformation of organic matter into gas and oil in the course of burial.

In addition to estimation of the amounts of oil and gas present in the formation and its organic carbon content, the parameters described above can also be used to recognise different source-rock horizons and to assess their oil-producing potential. The hydrocarbon quantity S_2 (oil + gas) resulting from cracking of the kerogen is equivalent to the total amount of oil and gas which the organic matter can still produce in the course of further maturation. This is called the residual hydrocarbon potential. It will be noted however that these values can be very adversely affected by adsorption of kerogen onto the mineral matrix.

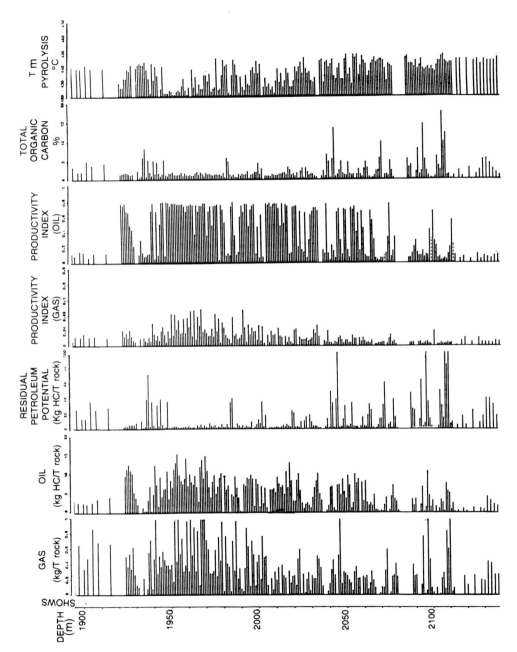

Fig. 104. — Oil show analyser — example of a geochemical log (Gabon).

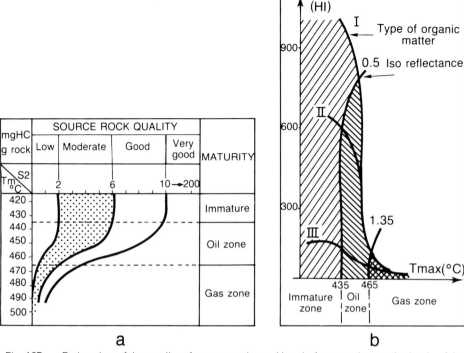

Fig. 105. — Estimation of the quality of source rocks and level of maturation on the basis of the relationship between S2 and Tm (a). Determination of the type of organic matter and its maturation using the relationship between HI and Tm (b).

The maturation state of the source rocks is estimated roughly on the basis of the Tm values (Fig. 105). It has been demonstrated that the temperature of the S_2 peak increases in relation to the level of maturity of the source rocks.

Figure 104 provides an example of an actual geochemical log obtained using the show analyser method (pyrolysis).

The origin of the organic matter within the source rocks can be defined roughly as : lacustrine (essentially type I), marine (type II) or continental (type III). The values of the hydrogen index (HI) plotted on a HI-Tm diagram (Figure 105b), contribute to the determination of these various origins.

Up to now this method has not been developed as a means for detecting zones of abnormal pressure. However any zone with a significant temperature and/or pressure anomaly is likely to be accompanied by a change in the maturation of the organic matter. Following up changes in the production index of clays and the maximum temperature Tm may perhaps make it possible to detect such changes. Although difficult to interpret in wildcats, these markers are currently in use in basins which have already been explored. The identification of undercompacted zones, which represent a perfect example of temperature and pressure anomalies, can therefore be envisaged. The fact that undercompacted shales are frequently also source rocks enables them to be recognised indirectly through an increase in total organic carbon.

The information provided by pyrolysis in the context of the detection of zones of abnormal pressure is not however unambiguous. Its use for this purpose alone cannot be justified at present in view of its high cost.

3.4.3.8. Nuclear magnetic resonance

The technique of nuclear magnetic resonance, which was discovered in the 1940's and has been used in wireline logs for some fifteen years, has only very recently been adapted to the analysis of cuttings on site in order to evaluate their porosity.

In order to explain the principle of this method it will be necessary to provide a reminder of some physical concepts :

Some nuclei have a spin with properties comparable to those of a kinetic moment. This spin has a corresponding magnetic moment.

To put it more simply, if the nucleus is regarded as a charged sphere, the charges turning around the axis give rise to a magnetic field. If the nucleus is placed in a magnetic field it behaves like a spinning top and precesses around the direction of the field.

The magnetic moments μ measured in nuclear magnetons differ :

— hydrogen = 2.7927 m · n
— carbon 13 = 0.7021 m · n
— sodium = 2.2161 m · n

Likewise the frequency of the precession is a function of μ and the applied magnetic field intensity, hydrogen having a very high frequency.

It is important to emphasise that some nuclei, in particular carbon 12 and oxygen 16, do not have any spin.

Thus NMR may effectively be used to detect hydrogen and thus water and hydrocarbons in porosity.

In practice nuclear magnetic resonance consists of sweeping the rock with a magnetic field at a fixed frequency using a polarising coil. The energy source of the coil is then cut and an electromotive force is induced in the coil by the precession of the hydrogen nuclei. The amplitude of the energy released is a measure of the number of hydrogen nuclei.

The value of the free fluid index is found by extrapolating the signal curve at the instant the energy is cut. It also depends on whether the fluid is water or hydrocarbon in nature.

The use on site of this new portable equipment should make it possible to determine the porosity of clays accurately and should thus favour the detection of undercompacted zones. It is also possible to envisage a quantitative application for evaluating pore pressure.

It should be noted that the method is subject to the same limitations as the measurement of shale density, including whether cuttings are representative, mud type and temperature, measurement procedures.

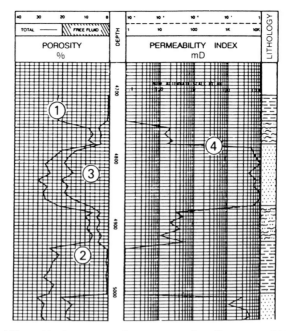

Fig. 106. — Nuclear magnetic resonance log (Courtesy of EXLOG).

Figure 106 shows a sample record from a nuclear magnetic resonance log, giving : total porosity (1), irreducible water saturation (2), the free fluid index (3) and an estimation of permeability (4).

3.4.3.9. Other methods

Numerous other methods, some of which are anecdotal, abound in the literature. FERTL (1976), in his work "Abnormal Formation Pressures", discusses the benefits of carrying out measurements on clay suspensions in distilled water and on the filtrate from these :

— resistivity of the suspension,

— filtration rate of the suspension,

— colour of the extracted filtrate (organic matter content),

— concentrations of HCO_3^- ions and other ions in the filtrate,

— redox potential and pH of the suspension

These methods are all difficult or time consuming. Results depend very largely on the nature of the mud and its interactions with the cuttings on their way to the surface. The main criticism is that these results are, in general, poorly argued, and presented like recipes based on a misunderstood or badly explained theory. For example the colour of the filtrate will not be representative of the organic matter content of clays if lignosulfonates are added to the mud.

Compared with techniques which are simpler to use, more reliable and above all easier to interpret, the usefulness of these methods appears negligible.

3.4.4. *CONCLUSION*

There are many methods for the detection of abnormally pressured zones while drilling, and they vary considerably in effectiveness. There is therefore a need to classify them on the basis of their reliability : reliable, moderately reliable, not very reliable. Lithology is an essential factor in interpretation, and must be linked to each of these parameters. Methods which are the subject of research are mentioned separately.

Detection parameter reliability	Real time methods	Delayed (lag time) methods
RELIABLE	Drilling rate Corrected "d" exponent (without wear factor) Normalised drilling rate Sigmalog Drag while making trips or connections Flow measurement Pit levels	Gas : — Connection gas — Background gas — Reservoir gas
MODERATELY RELIABLE	M.W.D. (penalised by the absence of a porosity log) Bottomhole settling of cuttings (resumption of drilling) Torque	Gas : - Gas show composi- tion - Trip gas Shale density Shale factor Pyrolysis Abundance and size of cut- tings Cuttings shape
NOT VERY RELIABLE	Pump pressure	Mud temperature
RESEARCH IN PROGRESS (Reliability to be established)		X-Ray diffraction Cuttings gas Nuclear magnetic resonance

This classification of the quality of detection methods must be seen as a generalisation. In fact, levels of reliability may vary in relation to geographical location (effect of lithological sequences, compaction, onshore/offshore location of wells, etc.)

Nigeria — 1980-1983
Comparative statistical study of the reliability of detection parameters

Well Name	Permit №	"d" Exponent	Background gas	Connection gas	Shale density	Shale Factor	Mud temperature	Kick	Mud logging co. A : Analysts B : Baroid G : Geoservices	Total depth	Top u/compac.
Odum - 1	OPL 465	1	1	4	4	7	6	yes	G	2 547	2 050
Toriye - 1	«	3.5	1	1	6	7	7	yes	G	3 063	2 720
Ofrima - 1	«	3	1	4	4	7	7	no	G	3 268	2 520
Ené - 1	OPL 458	6	3	2	4	7	6	yes	G	2 448	2 790
Aneka - 1	OPL 459	1	2	3	4	7	6	no	G	2 850	2 620
Otuo - 3D	OML 59	1	1	2	3	7	6	yes	G	2 850 TVD	2 760 TVD
Obogoro - 1	«	1	1	3	3	7	6	no	G	3 500	3 290
Ibewa - 2	OML 58	3	6	6	5	4	6	yes	B	4 433	3 700
Ibewa - 3	«	3	5	6	3	6	4	yes	A	4 507	3 670
Ihugbogo - E1	«	1	2	4	3	6	6	yes	A	4 206	3 840
Erema - W - 2	«	3	4	5	4	6	6	yes	A	4 187	3 590
Jatumi - 2	OML 57	2.5	2	2	5	5	6	no	G	3 295	3 100
Okpoko - S - 1	«	6	6	6	6	6	6	yes	G	3 268	3 200
Okpoko - S - 1b	«	4	3	4	4	5	6	no	G	3 691 TVD	3 030 TVD
Obodugwa - 1	OML 56	2	4	5	6	4	4	no	B	4 356	3 500
Obodugwa - 2	«	2	3	6	4	5	5	no	B	4 199	3 600
Obodugwa - 3	«	2	3	4	6	5	5	no	B	4 108	3 460
Umusadege - N2	«	2	6	6	4	5	5	yes	B	4 052	3 420
Igbuku - 1 + ST	«	2	4	6	3	3	6	yes	B	4 196	3 000
Igbolo 1	«	4	6	6	4	5	6	yes	B	3 434	3 090

Summary								
	Average reliability		2.6	3.2	4.2	4.2	4.7	5.6
	Reliability per permit	OPL 465	2.5	1	3	4.6	—	6
		OPL 458 459	3.5	2.5	2	3.5	—	6
		OML 59	1	1	2.5	3	—	6
		OML 58	2.5	4.3	5.3	3.8	5.5	5.5
		OML 57	4.1	3.7	4	5.3	5.3	6
		OML 56	2.3	4.3	5.5	4.5	4.5	5.2

COMPARISON ·SCALE

1 Very reliable quantitatively
2 Very reliable qualitatively
3 Reliable
4 Moderately reliable
5 Not very reliable
6 Inconclusive
7 Not used
(excluded from averages)

It should always be remembered that the origins of overpressure (e.g. undercompaction) may not be present at the vertical of the borehole (laterally induced pressures). In such cases compaction indicators will be of no help whereas differential pressure indicators, notably gas measurements, will be essential.

By way of example, a statistical study performed in Nigeria on 20 wells drilled between 1980 and 1983, in which the effectiveness of methods is noted, is given below. On the whole it is found that the "d" exponent and background gas are the best techniques, while mud temperature and shale factor are the least effective. For information, OPL's 458 and 465 are offshore, OPL 459 and OML 59 close offshore, OML 57 in the swamp and OML's 56 and 58 onshore.

3.5. METHODS AFTER DRILLING

This chapter describes the various methods for detecting abnormal pressure which are used after drilling. Some of the methods can now be used while drilling (M.W.D.) and most of them can intervene before the end of a well during intermediate logging runs or tests.

3.5.1. *WIRELINE LOGS*

Interpreting the data from wireline logs affected by changes in shale porosity can confirm or define the existence of abnormal pressure zones following detection by real time measurements. New techniques, gamma ray spectrometry for example, are opening the way to investigation of abnormal pressure due to causes other than undercompaction.

We shall now discuss the main logging methods, some of which border on the domain of M.W.D., and look at the information which can be provided by automatic data processing.

3.5.1.1. Resistivity/conductivity

Shale resistivity is one of the oldest methods for detecting abnormal pressure.

As rock matrices have very low conductivity, recorded resistivity depends on their porosity, the nature of the fluid contained in the pores and its dissolved salt content.

Under normal compaction conditions, a unit increase in shale resistivity with depth corresponds to unit reduction in porosity under the effect of the weight of overlying sediments (at a given fluid resistivity).

Entry into an undercompacted zone is revealed by a fall in resistivity due essentially to a relative increase in porosity (Figure 107).

Nevertheless factors other than porosity affect measurements of formation resistivity and can mask changes due to compaction.

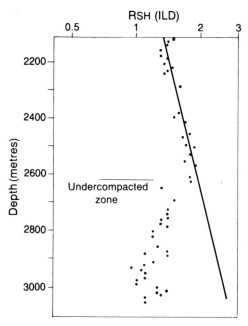

Fig. 107. — Changes in shale resistivity with depth in an undercompacted zone (offshore Nigeria).

As we have already seen, entry into an undercompacted zone is frequently accompanied by a fall in salinity. It has the opposite effect of the porosity increase but is an order of magnitude smaller.

— temperature : this increases with depth, resulting in a decrease in resistivity for water of a given salinity
— presence of hydrocarbons : the presence of hydrocarbons in the formation pore space considerably increases its resistivity :

$$\text{oil} : 10^6 \text{ to } 10^9 \text{ } \Omega/m^2/m$$
$$\text{gas} : \text{infinity}$$

— organic matter : the presence of large proportions of organic matter also increases resistivity. This factor may attenuate the effects of undercompaction (such zones are frequently rich in organic matter).
— lithology : even a very small difference in lithology (slightly silty clay in comparison with a pure clay) can cause an error when using resistivity to determine the normal compaction trend. In cases of considerable undercompaction (Gulf of Guinea) the effect of minor lithological changes becomes negligible.
— wash-outs : increase in hole diameter due to caving may also give rise to an error in measuring shale resistivity. Examining the caliper log should help to correct this problem.

Not all devices give the same resistivity measurement for clay/shale lithology. This can be explained by the anisotropy of the matrix, giving different vertical and horizontal resistivities. Because of the way the induction log operates, it detects mainly horizontal resistivity and so yields the lowest value.

The resistivities to be used for the preparation of compaction plots should be those obtained using techniques with a deep investigation. The method most frequently used in the past was the 16-inch short-normal currently adopted by M.W.D. techniques. The most suitable conventional device for our purposes is the deep induction tool.

Figure 108 is an example of an induction recording in an undercompacted zone. The fall in deep resistivity is made more obvious by amplifying the trace. Conductivity logs may also be used to reveal compaction anomalies, as conductivity yields more detail in low resistance zones. The quantitative interpretation of resistivity variations is sometimes made difficult by the existence of non-linear compaction trends in the overlying hydrostatic series.

3.5.1.2. Sonic

The sonic logging tool measures the sound wave transit time per foot in a vertical direction in the vicinity of the borehole, that is to say the reciprocal of the longitudinal sonic velocity in the formation.

Results are expressed as transit time over a given interval expressed in microseconds per foot.

In a porous argilo-detritic series, sonic transit time may be derived approximately using the following formula :

$$\Delta t = \phi \, \Delta t_f + (1 - \phi) \, \Delta t_m$$

where Δt = measured transit time (μsec/foot)
Δt_f = transit time in the fluid (μsec/foot)
Δt_m = transit time in the matrix (μsec/foot)
ϕ = porosity

Transit times are faster in the matrix (of the order of 40-55 μsec/foot) than in fluids (200 μsec/foot for water) :

Medium	Δt (μsec/foot)
Methane	600 to 700
Water	170 to 220
Oil	238
Quartz	55.5
Calcite	46.5
Dolomite	38.5 to 44
Rock salt	67
Anhydrite	50
Steel (casing)	58

For a given lithology, transit time depends on porosity (except when free gas is present). Since porosity decreases with depth, so does the sonic transit time.

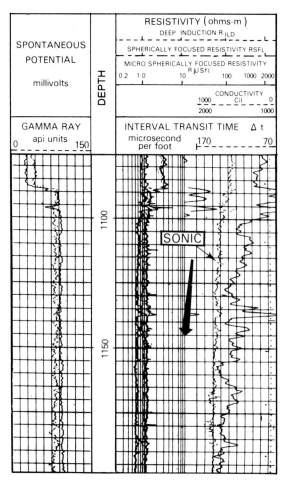

Fig. 108. — An example of sonic, resistivity and conductivity responses in an undercompacted clay (Niger Delta, Cameroon).

Porosity is obtained using the following formula :

$$\phi = \frac{\Delta t - \Delta t_m}{\Delta t_f - \Delta t_m}$$

Thus in the case of clays the sonic logging tool provides an excellent way of assessing compaction qualitatively and quantitatively (Fig.108).

The above example shows an increase of 10 µsec/foot between 1 100 and 1 170 m corresponding to undercompaction. This is also marked by the deep induction and by the conductivity.

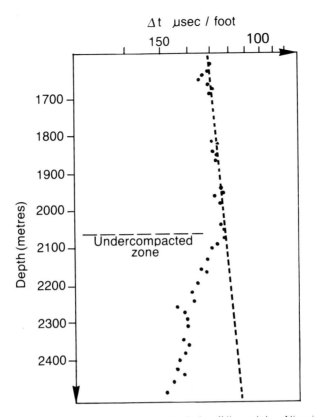

Fig. 109. — Example of a Δt plot in shales (Niger delta, Nigeria).

A plot of transit times for selected clay levels on a logarithmic scale against linear depth shows compaction anomalies more clearly (Fig.109).

If lithology is constant, the normal compaction trend can be taken to be a straight line. Lithological variations in the clay (changes in silt or carbonate contents) have the effect of reducing transit times and thus of causing a shift in compaction trends.

Every clay facies has its own compaction trend parallel to the trend line drawn for pure clays. The equation for the trend line can be defined regionally on the basis of statistical data. In a clay series with variable facies this assists identification of the undercompacted zone (Figure 110a).

Within an undercompacted series, clays of different facies may also be revealed (Figure 110b).

We have already examined the effect which a number of factors can have on the slope of the normal compaction trend (section 3.1). Recent studies at Elf Aquitaine have demonstrated the effect of the sand-shale ratio in particular on this trend.

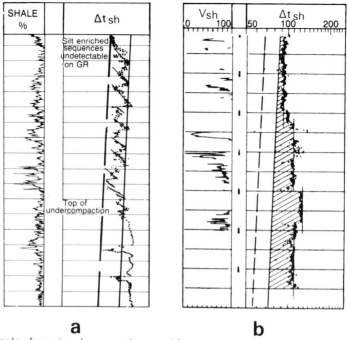

Fig. 110. — Example of a regional compaction trend (computer processed), silty sequences (a) — facies variations in clays within an undercompacted zone (b) (Chiarelli *et al.*, 1973, courtesy of IFP).

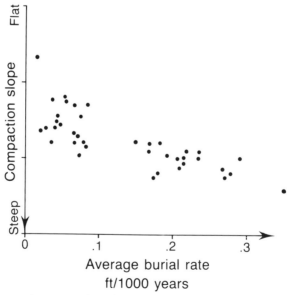

Fig. 111. — Changes in compaction slope in relation to rate of burial — North Canada (Magara, 1978, courtesy of Elsevier ed.).

MAGARA (1978) has established a relationship between the rate of burial and the slope (Figure 111).

Sonic logging, which allows the determination of shale porosity, is one of the most reliable logging techniques for identifying undercompacted zones. Its use requires a normal compaction trend to be established, and this should preferably be determined on a regional scale.

3.5.1.3. Density

If a source of gamma rays is applied to the wall of a borehole, an interaction takes place between the gamma rays and the material due essentially to gamma/electron collisions. The energy of the incident photon is partly transmitted to the electron ejected from an atom. The density tool measures the strength of the diffused gamma radiation. The number of electrons in atoms are approximately proportional to their density, collisions are therefore more numerous the denser the material.

Gamma ray attenuation is thus directly dependent on formation bulk density. If the density of a matrix is known, its porosity can be determined :

so that :
$$\rho_b = (1 - \phi) \rho_m + \phi \rho_f$$
$$\phi = \frac{\rho_m - \rho_b}{\rho_m - \rho_f}$$

where : ρ_m = matrix density
  ρ_b = measured bulk density
  ρ_f = fluid density.

For our purposes, reading off shale bulk densities from the logs gives an indication of compaction, assuming matrix and fluid densities to be constant.

Figure 112a is an example of an FDC (Formation Density Compensated - Schlumberger) trace in an undercompacted zone showing a progressive fall in density between 1 100 and 1 260 m (from 2.25 to 2.10).

Figure 112b is a logarithmic plot of clay densities revealing the top of undercompaction.

As observed with the logarithmic plots of the majority of other logging tools seen earlier, the normal compaction trend obtained using densities is effectively linear.

Measurements obtained with density tools are frequently affected by factors other than lithology and porosity (hydrocarbons, particularly gas). The shallow investigation depth of the tool makes it sensitive to the state of the borehole walls : caving, cake thickness, clay hydration. In the upper part of Figure 112b the shift in density values can be attributed to hydration of montmorillonite-rich clays.

Although also affected by these measurement anomalies, but to a lesser extent, the sonic log usually gives more satisfactory results than the density log.

The normal compaction trend for density is often difficult to establish for the simple reason that density logs are rarely run in top hole.

173

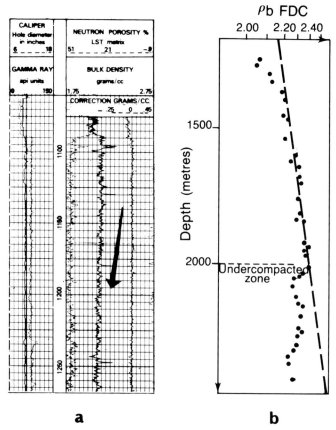

a   **b**

Fig. 112. — Changes in clay bulk density in an undercompacted zone. FDC trace — Cameroon ; (a)
— manual plot of FDC data-Nigeria (b).

3.5.1.4. Neutron porosity

In essence, the neutron method measures the amount of hydrogen in a given volume of formation. High energy neutrons are continuously emitted by a radioactive source. Their energy is reduced when they collide with the material in the matrix. The presence of hydrogen nuclei of the same mass as the neutrons causes maximum energy loss.

Measurement of the amount of hydrogen provides an approximation to the porosity of the formation.

The neutron log does not provide a measure of total porosity because neutrons do not distinguish whether protons belong to free or adsorbed liquids. Protons in shales (and in hydrated salts such as gypsum) are therefore included. For this reason it is better to speak of the neutron hydrogen index than neutron porosity. A clay frequently has a total neutron porosity measurement of 40 to 50 %.

The neutron log is not used for a qualitative or quantitative analysis of undercompaction. It may however contribute to understanding the origins of abnormal pressure. Clays rich in montmorillonite cause a fall in neutron energy, and high porosity is recorded because of their high adsorbed water content. Illites, which have a low adsorbed water content, have very much lower neutron porosities.

A rise in neutron porosity when entering an abnormally pressured zone may indicate increased levels of montmorillonite which can be associated with undercompaction.

Abnormal pressure arising from the dehydration of montmorillonites may be reflected either in stable neutron porosity, because the free water is trapped in the pore space, or in reduced porosity if some of the free water has been expelled. When interpreting data, any gas effect which would reduce the normal neutron value must also be considered.

However, it is usually found that neutron data are difficult to interpret for the purposes of identifying undercompacted zones.

3.5.1.5. Gamma-ray/spectrometry

The gamma-ray logging tool measures gamma rays emitted naturally by the formation. The concentration of radioactive minerals varies with lithology. For clay facies, the radioactivity depends mainly on the presence of potassium-bearing minerals (illite, interstratified clays). Where these are uniformly concentrated within a clay series, observed variations will be due to changes in porosity only. The arrival of M.W.D., which has a smaller number of parameters available, motivates a closer interpretation of the gamma-ray log. Reduced radioactivity is observed in some undercompacted zones (see Figure 49). Interpretation is difficult, and must take account of lithological changes (sequences) as well as other factors which influence gamma-ray readings : hole diameter, mud type, etc. It is thus necessary to use a corrected gamma-ray reading, though sometimes such corrections are greater in magnitude than the variations in the phenomenon under investigation. Some authors (ZOELLER, 1984) suggest that gamma-ray data might even be used for quantitative evaluation of pressure !

In addition to measuring total radioactivity, a radiation spectrum may be used to find the concentrations of thorium, potassium and uranium contributing to the total gamma-ray reading.

Thorium and potassium concentrations in clays can vary. According to work by HASSAN & HOSSIN (1975) the Th/K ratio is a good indicator of the nature of the clay mineral (Fig. 113) and can be used to adjust regional trends, for example by eliminating silts which are more radioactive than pure clays, or zones containing fine volcanic material (zircons), where total gamma ray readings can be misleading.

Th/K cross-plots using the spectrometry log on well-defined clay horizons can in theory be used to identify clay minerals or, more particularly, to follow their changes (Figure 114). This diagram, which is too schematic, does not cover all situations encountered, for example pyrophyllite and allevardite clay series have no radioactive response.

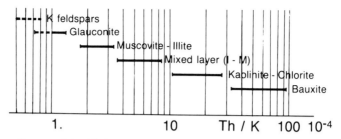

Fig. 113. — Range of the Th/K ratio in some minerals (after Hassan & Hossin, 1975).

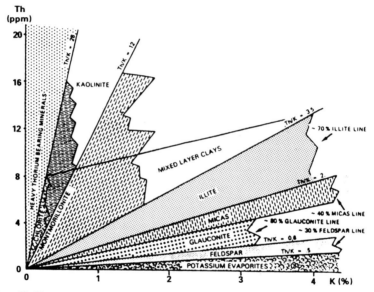

Fig. 114. — Th/K crossplot which can be used for mineralogical interpretation based on the spectrometry log (Courtesy of Schlumberger).

However, in studying smectite transformations, the information used can be viewed in conjunction with the shale factor and thus contribute to an understanding of the causes of abnormal pressure (section 2.3).

The spectral method undoubtedly provides much more complete information than an overall measurement of radioactivity. However, the inherent limits to the interpretation of gamma-ray tools explained above restrict their use to the understanding of the mechanism of abnormal pressure.

3.5.1.6 Conclusion

Detection methods used at the end of a drilling phase are not only able to give an indication of the presence and magnitude of abnormal pressure, but also contribute to the programme for the following phase.

Figure 115 schematically summarizes the responses of the various wireline logs when entering an undercompacted zone.

It should by now be clear to the reader that log readings in the overlying normally compacted series are fundamental to the interpretation of readings in the abnormally pressured zone. This should be borne in mind when preparing logging programmes.

Care must be taken when choosing and calibrating logging tools. In addition certain precautions are essential when handling and interpreting data :

— *avoid using shallow data : hydration phenomena, caving, etc.*

— *ignore beds too thin for the true logging value to be attained due to limitations in vertical tool resolution — this leads to falsification of the trends.*

— *various other phenomena can affect the measurement validity :*
  - *organic matter and gas content of the clays,*
  - *reduced resistivity in the vicinity of salt domes,*
  - *well conditions and the conditions under which measurements were recorded (hole diameter, tool calibration etc).*

As was noted in previous chapters, the efficiency of overpressure analysis depends largely on a judicious selection of data and the way they are plotted.

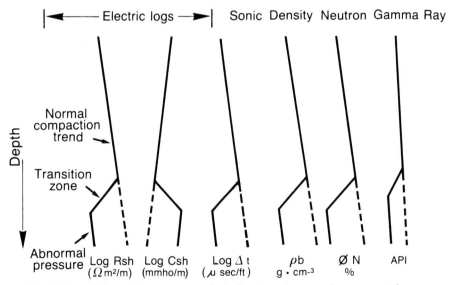

Fig. 114. — Schematic responses of wireline logs in an undercompacted zone
(modified from Fertl & Timko, 1971).

3.5.2. *LOG-DATA PROCESSING BY COMPUTER*

Clay porosity, or any parameter associated with it, can be plotted in relation to depth on a semi-logarithmic scale. We have seen that a straight line, known as the trend, is obtained where compaction is normal. The points corresponding to undercompacted clays are offset to a greater or lesser extent from this trend. Resistivity, acoustic velocity and density are some of the parameters associated with porosity, which is why they are used for investigating compaction anomalies. The greater the porosity of a sediment, the greater will be the acoustic velocity, and the lower the density and resistivity (assuming the fluid type and matrix characteristics remain the same).

Most companies possess programs which automatically process log data to produce compaction profiles.

It can sometimes be difficult to evaluate the lithology from an automatic trace, hindering correct determination of the compaction.

Because every facies has its own normal compaction trend, as many plots as possible need to be taken in order to avoid a "lithological stumbling block". Only a computer can cope with this.

3.5.3. *VELOCITY SURVEY OR CHECKSHOT*

The principle of a velocity survey, or "checkshot", consists of lowering a geophone, which is brought to a halt at various preselected depths in turn, down the well. Shots are fired on the surface, at a point a short distance away from the well. The mean velocity between the firing point and the geophone depends on the distance L separating these two points, so that :

$$V_G = \frac{L}{t}$$

Corrections are applied to times (t) in order to obtain the times which would have been measured directly if the shot had been fired at the rotary table.

The various time/depth pairs can be used to plot the following :

— time/depth graph
— mean velocity graph
— interval velocity graph

This is most commonly done at the end of a well, but it may be carried out while drilling, for example when intermediate wireline logs are being run. In addition to its usefulness in calibrating the sonic log, the velocity survey can be used to correct surface seismic information by establishing the true time/depth relationship. Thus a velocity log which is run before the objective is reached will have the advantage that the depth of any given seismic anomaly, such as that assumed to represent the top of the undercompacted zone, can be redefined.

178

3.5.4. *VERTICAL SEISMIC PROFILE*

In comparison with the velocity survey, the VSP tool is fixed at regularly spaced levels (20 to 30 m). The complete seismic trace is recorded.

In addition to direct waves, i.e. the same as those obtained by a checkshot, the events recorded are of two types :

— downward travelling waves which reach the geophone from above. On the seismic section these arrivals appear parallel with the direct arrivals,

— upward travelling waves, which reach the geophone from below. On the seismic section the slope of these is opposite to that of the downward arrivals.

By filtering and subtracting the downward-travelling wave field from the upward-travelling wave field it is possible to record the latter only, giving a seismic signature which is of better quality than that which is obtained at the surface, particularly below the bottom of the hole.

In addition to the useful features which it shares with the velocity survey, VSP can be used to correlate the traces obtained with those obtained from surface seismography.

4. — QUANTITATIVE PRESSURE EVALUATION

4.1. FORMATION PRESSURE EVALUATION

Several methods of evaluating formation pressure are available for use before, during and after drilling. Only the detection methods below can be used to evaluate pressure quantitatively :
- formation tests, which provide a direct measurement of the pressure,
- seismic interval velocities,
- drilling rate : "d" exponent, Sigmalog, normalised drilling rate,
- shale density,
- gas shows,
- kicks/mud losses : mud flow measurements, pit levels,
- wireline logs : resistivity/conductivity, sonic, density.

Most methods of evaluation are based on the principle of comparing the undercompacted clays with a normal compaction state, which necessarily means that the normal compaction trend must be obtained for the parameter investigated. This approach is based on the geostatic origin of the pressure. Pressures are calculated on the assumption that there is a direct relationship between the porosity anomaly and the pressure anomaly. The deduced pressure is considered to be close to that of a confined reservoir located at the depth in question.

None of the methods available should be universally applied. Results vary in their degree of accuracy, and must be used with caution. Accuracy and method selection are likely to improve as more wells are drilled in a given region.

The following methods are used to evaluate pressure. Formation tests, which are the only means of measuring it, are dealt with in section 4.1.7.

4.1.1. *EQUIVALENT DEPTH METHOD*

Applications : interval velocities, "d" exponent, shale density, resistivity/conductivity, sonic, density logs and any direct or indirect measurements of clay porosity.

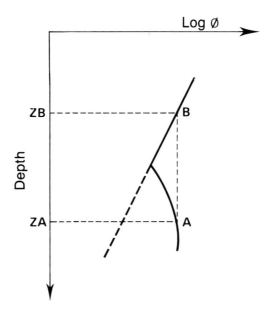

Fig. 116. — Principle of the equivalent depth method.

Principle : every point A in an undercompacted clay is associated with a normally compacted point B. The compaction at point A is identical to that at point B (Fig. 116).

The depth of point B (ZB) is called the equivalent depth, or sometimes the isolation depth. The fluid contained within the pores of clay A has been subjected to all the geostatic load in the course of burial from ZB to ZA.

Using Terzaghi's formula (see section 1.3) :

$$S = \sigma + P$$

The matrix stress σ transmitted by grain to grain contact is identical at A and B.

Knowing the overburden pressure S_B and the normal pore pressure at B (P_B), σ_B can be calculated.

$$\sigma_B = S_B - P_B \qquad (4.1)$$

as $\sigma_B = \sigma_A$

Knowing the overburden pressure S_A at A, the pore pressure at A is obtained.

$$P_A = S_A - \sigma_B \qquad (4.2)$$

then, by eliminating σ_A and σ_B :

$$P_A = P_B + (S_A - S_B) \qquad (4.1) \text{ and } (4.2)$$

Example :

A = 3 500 m, B = 2 500 m, normal pressure gradient = 1.06, overburden gradients :
2.20 at B, 2.26 at A.

$$P_B = \frac{ZB}{10} \times 1.06$$

$P_B = 265 \text{ kg/cm}^2$

$$S_B = \frac{ZB}{10} \times 2.20$$

$S_B = 550 \text{ kg/cm}^2$

$$S_A = \frac{ZA}{10} \times 2.26$$

$S_A = 791 \text{ kg/cm}^2$

$$P_A = 265 + (791 - 550) = 506 \text{ kg/cm}^2$$

Thus, for the equilibrium density :

$$d_{eql} = \frac{P_A}{ZA} \cdot 10$$

$$d_{eql} = 1.45 \text{ g} \cdot \text{cm}^{-3}$$

The formula to be used at the wellsite, when the overburden gradient is known, is :

$$d_{eql}A = GG_A - \frac{ZB}{ZA} (GG_B - d_{eql} B) \qquad (4.3)$$

where
$d_{eql} A$ = equilibrium density at A
$d_{eql} B$ = equilibrium density at B
ZB = equivalent depth
ZA = depth of the undercompacted clay
GG_A = overburden gradient at A
GG_B = overburden gradient at B.

Calculation of the overburden gradient is described below (see section 4.2).

If data for calculation of the overburden gradient are not available, an average overburden gradient may be used. The value normally taken is 2.31 (1 psi/foot), which corresponds to an average established for the Gulf Coast. Although this value produces only a small error in the case of onshore wells, it should not be used offshore if at all possible, particularly where the water is deep and the well is shallow.

When the normal pressure gradient is not known an average value of 1.05 may be substituted for it.

Simplified formula for constant gradients (d_{eql} B = 1.05, GG = 2.31) :

$$d_{eql}A = 2.31 - \frac{ZB}{ZA} (2.31 - 1.05) \qquad (4.4)$$

giving :
$$d_{eql}A = 2.31 - 1.26 \frac{ZB}{ZA} \qquad (4.5)$$

183

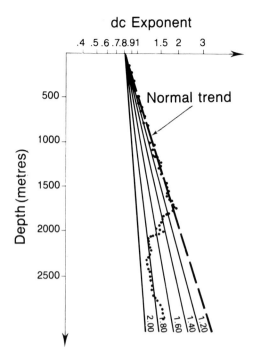

Fig. 117. — Example of the set of isodensity lines obtained by the equivalent depth method ("d" exponent).

If we substitute the depths used in the previous example, we obtain : d_{eql} A = 1.41.

When the compaction trend has been established, pressure calculations can be made easier by drawing a set of isodensity lines, from which equilibrium densities can be read off directly (Figure 117).

Establishing isodensity lines (Figure 118)

— extend the normal compaction trend (XY) to the depth origin (X)

— choose a point (B) located on the normal compaction trend

— for a selected value of d_{eql}A calculate depth A using the following formula derived from (4.5) :

$$ZA = \frac{1.26\ ZB}{2.31 - d_{eql}A} \qquad (4.6)$$

— position point A on the vertical from B, then draw a straight line XZ passing through A.

184

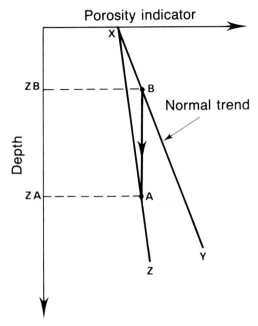

Fig. 118. — Drawing the set of isodensity lines.

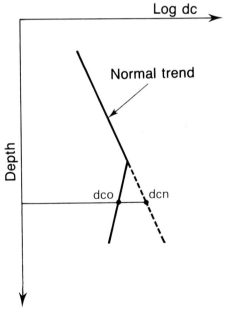

Fig. 119. — Ratio method : principle ("d" exponent).

It should be noted that if the overburden gradient and the normal pressure gradient are known it is better that these should be substituted for the constants used in formula (4.6).

The equivalent depth method may be used regardless of whether the porosity parameter concerned is represented arithmetically or logarithmically.

4.1.2. *RATIO METHOD*

Applications : "d" exponent, shale density, sonic log, resistivity/conductivity, and density log.

Principle : the difference between observed values for the compaction parameter and the normal parameter extrapolated to the same depth is proportional to the increase in pressure (Figure 119).

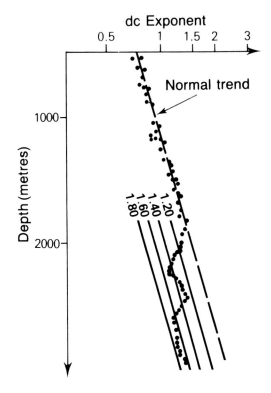

Fig. 120. — Example of a set of isodensity lines.

The equilibrium density is obtained using the following formula :

$$d_{eql} = d_{eql\,n} \cdot \frac{dc_n}{dc_o}$$

where $d_{eql\,n}$ = normal equilibrium density
dc_n = "normal" "d" exponent
dc_o = observed "d" exponent

A set of isodensity lines can be drawn using the following formula (Figure 120) so that the equilibrium densities can be read off directly.

$$dc_o = dc_n \cdot \frac{d_{eql\,n}}{d_{eql}}$$

Establishing isodensity lines (Figure 121)
— take a point (A) located on the normal compaction trend XY
— calculate the value of dc which would be observed at point A for a given equilibrium density
— using this value (B) draw a straight line X'Y' parallel to XY. This represents the gradient of the selected equilibrium density.

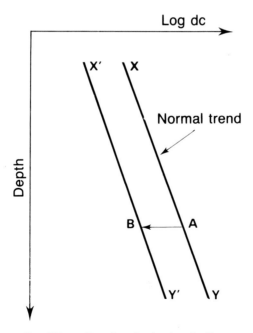

Fig. 121. — Drawing the isodensity lines.

Example : $dc_n = 1.80$, $d_{eql\,n} = 1.05$

$$dc_o = 1.80 \, \frac{1.50}{d_{eql}}$$

to draw the isodensity line $d_{eql} = 1.20$,

$$dc_o = 1.80 \, \frac{1.50}{1.20} = 1.58$$

The ratios method is easy and very widely used. However, because it is empirical, the results obtained are not always satisfactory. Adjustment of the calculations on the basis of measurements (RFT, tests) can appreciably improve the results of the method with the introduction of a correction coefficient (c) :

so that : $$d_{eql} = c \cdot d_{eql\,n} \cdot \frac{dc_n}{dc_o}$$

where c = the correction coefficient

Example : calculated $d_{eql} = 1.25$
RFT d_{eql} $= 1.35$

$$c = \frac{1.35}{1.25} = 1.08$$

This correction coefficient remains applicable as long as the origin and the causes maintaining the abnormal pressure remain constant for the unit in question.

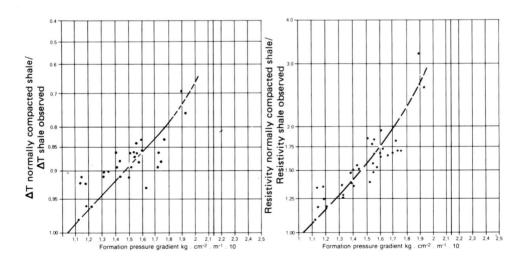

Fig. 122. — Δt and resistivity ratio/pressure gradient charts (Niger Delta).

In the case of the "d" exponent, an estimate can only be made if a linear relationship is assumed between compaction and formation pressure, as both these factors affect drilling rate.

Ratio/pressure graphs may be compiled when a large amount of data is available on pressure and porosity parameters (Figure 122a and b).

The above two charts are examples from wells drilled in the Gulf of Guinea and show the relationship between resistivity or transit time ratios and measured pressures. Note the considerable scatter of the points in relation to the mean value established by the least squares method. For example, as far as Δt is concerned, a ratio of 0.86 is equivalent to a pressure gradient of between 1.4 and 1.73.

Other relationships established by HOTTMAN & JOHNSON (1965) and FERTL & TIMKO (1970) in the Gulf Coast show less scattered results for a smaller set of data (Figures 123a and 123b).

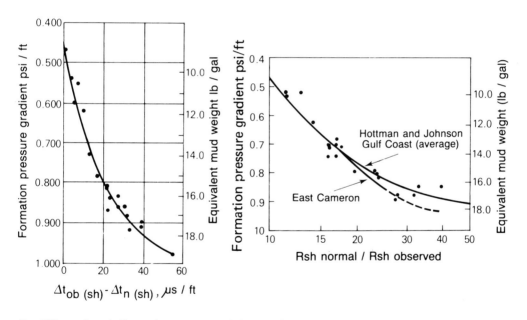

Fig. 123a. — Δt ratio/formation pressure relationship (Hottman & Johnson, 1965, courtesy of SPE of AIME).

Fig. 123b. — Resistivity ratio/formation pressure relationship (modified from Fertl & Timko, 1972-73, courtesy of SPE of AIME).

Because of their statistical nature, these ratios cannot be reliably applied except in zones which have already been well explored, and preferably in sectors of limited extent.

Figure 124, developed by analysis of sonic ratios and pressure measurements on 60 wells from different regions, can be used to evaluate pore pressure from sonic ratio as a function of the sonic trend slope encountered (M). This plot covers the normal range of slopes observed. If the slope is not known with such precision the average slope (M = $-2.75 \cdot 10^{-4}$) should be used. In this case the error could however reach plus or minus 0.4 g. cm^{-3} d_{eql}. This figure clearly demonstrates the importance of slope definition for pore pressure analysis.

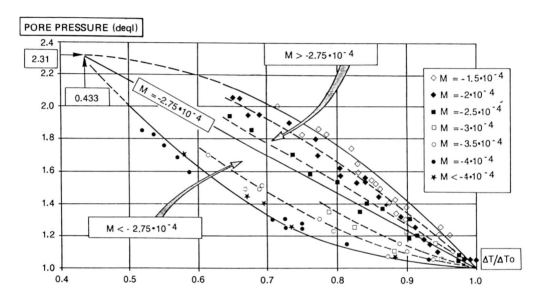

Fig. 124. — Δt ratio/formation pressure relationship as a function of trend slope (M).

4.1.3. *THE EATON METHOD*

Application : interval velocities, "d" exponent, resistivity/conductivity, sonic. It may also be extended to : shale density, log density.

Principle : the relationship between the observed parameter/normal parameter ratio and the formation pressure depends on changes in the overburden gradient.

EATON (1972) established the following empirical relationship from real data :

$$P = GG - 0.535 \left(\frac{R_{sh} \text{ observed}}{R_{sh} \text{ normal}}\right)^{1.5} \qquad 4.7$$

where : P = formation pressure gradient (psi/foot)
  GG = overburden gradient (psi/foot)
  R_{sh} = shale resistivity

According to TERZAGHI & PECK (1948) (see paragraph 1.3) :

$$\sigma = S - P \tag{4.8}$$

The normal matrix stress of 0.535 psi/foot (1.24 kg $\cdot$ cm^{-2} $\cdot$ m^{-1}. 10) only applies to a constant overburden gradient of 1 psi/foot (2.31 kg $\cdot$ cm^{-2} $\cdot$ m^{-1} $\cdot$ 10) and a normal hydrostatic pressure gradient of 0.465 psi/foot (1.07 kg $\cdot$ cm^{-2} $\cdot$ m^{-1} $\cdot$ 10).

Equations (4.7) and (4.8) are combined in order to take changes in both these factors into account :

$$P = GG - (GG - P_n) \left(\frac{R_{sh}\ observed}{R_{sh}\ normal}\right)^{1.5} \tag{4.9}$$

where P_n = normal formation pressure gradient.

EATON performed additional studies based on a large amount of data. These resulted in a modification to the exponent in his formula in 1975 :

$$P = GG - (GG - P_n) \left(\frac{R_{sh}\ observed}{R_{sh}\ normal}\right)^{1.2} \qquad \text{metric or US units}$$

Other relationships have been obtained for conductivity, the "d" exponent and the sonic log :

conductivity :

$$P = GG - (GG - P_n) \left(\frac{C\ normal}{C\ observed}\right)^{1.2}$$

"d" exponent :

$$P = GG - (GG - P_n) \left(\frac{d_c\ observed}{d_c\ normal}\right)^{1.2}$$

Δ_t sonic :

$$P = GG - (GG - P_n) \left(\frac{\Delta_t\ normal}{\Delta_t\ observed}\right)^{3.0}$$

Specimen calculation (for resistivity) :
GG = 2.26 kg $\cdot$ cm^{-2} $\cdot$ m^{-1} $\cdot$ 10
P_n = 1.07 kg $\cdot$ cm^{-2} $\cdot$ m^{-1} $\cdot$ 10
Observed R_{sh} = 0.68 ohm $\cdot$ m
Normal R_{sh} = 3.50 ohm $\cdot$ m

$$P = 2.26 - (2.26 - 1.07) \left(\frac{0.68}{3.5}\right)^{1.2} = 2.09\ kg \cdot cm^{-2} \cdot m^{-1} \cdot 10$$

The exponents defined above are sufficiently reliable for widespread use in exploration wells. However, if a sufficient number of pressure data are available they can be validated or corrected on a regional basis.

Eaton's method is undoubtedly the most widely used at the present time, even though it requires knowledge of the local overburden gradient.

If no data-processing equipment is available on site Eaton suggests that graphs should be used to evaluate the ratios (Figure 125a and b) and the formation pressure gradient (Figure 126), the normal pressure gradient being known or estimated.

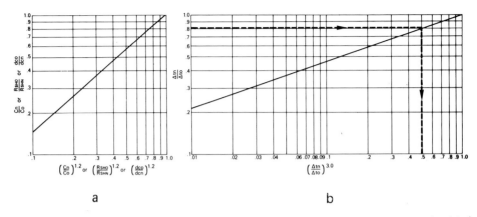

a   b

Fig. 125a). — Calculation of a ratio with an exponent of 1.2 ("d" exponent, resistivity, conductivity). (a)
— Calculation of a ratio with an exponent of 3.0 (transit time Δt) (b)
(from Eaton, 1976, courtesy of WORLD OIL).

Example of its use :

Observed $\Delta t = 100 \, \mu sec/ft$, normal $\Delta t = 80 \, \mu sec/ft$, overburden gradient $= 2.25$ $kg \cdot cm^{-2} \cdot m^{-1} \cdot 10$, normal pressure gradient $= 1.05 \, kg \cdot cm^{-2} \cdot m^{-1} \cdot 10$.

$$\frac{\Delta t_n}{\Delta t_o} = 0.8$$

From graph 125b we obtain :

$$\left(\frac{\Delta t_n}{\Delta t_o}\right)^{3.0} = 0.5$$

On graph 126 draw a horizontal line corresponding to :

$$\left(\frac{\Delta t_n}{\Delta t_o}\right)^{3.0} = 0.5$$

as far as GG = 2.25. Draw a vertical line from this point to the gradient axis.

Formation pressure gradient $= 1.65 \, kg \cdot cm^{-2} \cdot m^{-1} \cdot 10$.

Graphs of the type in Figure 126 must be prepared for every value of the normal pressure gradient. The preparation of isodensity lines on site may provide a more visual aid to evaluation of the "d" exponent.

192

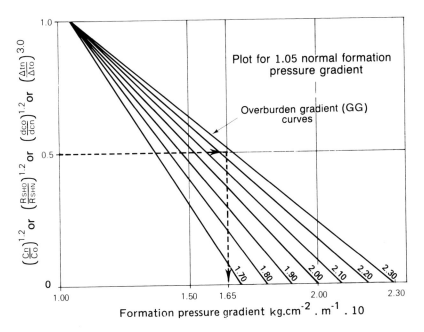

Fig. 126. — Chart allowing the formation pressure gradient to be read directly from the overburden gradient and ratio (modified from Eaton, 1972).

Establishing isodensity lines : for the "d" exponent
  — determination of the normal compaction trend
  — calculation of the theoretical values of the observed "d" exponent for different values of the pressure gradient P using Eaton's formula :

$$\text{dc observed} = \left({}^{1.2}\sqrt{\frac{GG - P}{GG - P_n}} \right) \times \text{dc normal}$$

By repeating the process at several intervals of 100-500 m, a set of isodensity lines comparable to that shown in the figure below can be obtained :

Specimen calculation (Figure 127) ; drawing the 1.20 isodensity line :

Data :

depth	1 000 m
normal "dc"	0.85
overburden gradient GG	1.95
normal equilibrium density	1.00

193

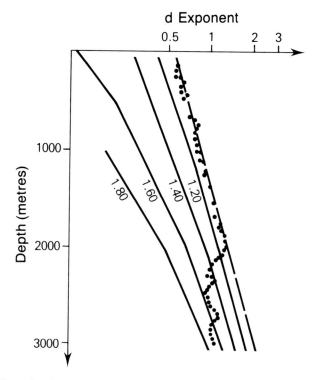

d Exponent

0.5   1   2   3

Depth (metres)

1000

2000

3000

1.80   1.60   1.40   1.20

Fig. 127. — Example of a set of isodensity lines for "d" exponent based on Eaton's formula.

$$dc_o = \left(1.2 \sqrt{\frac{S - P}{S - P_n}}\right) dc_n$$

$$dc_o = \left(1.2 \sqrt{\frac{1.95 - 1.20}{1.95 - 1.00}}\right) \cdot 0.85$$

$$dc_o = 0.70$$

at 1 500 m, normal "dc" = 1.05, GG = 2.04, P_n = 1.00

$$dc_o = \left(1.2 \sqrt{\frac{2.04 - 1.20}{2.04 - 1.00}}\right) \cdot 1.05$$

$$dc_o = 0.88$$

at 2 000 m : normal "dc" = 1.30, GG = 2.15, P_n = 1.00

$$dc_o = \left(1.2 \sqrt{\frac{2.15 - 1.20}{2.15 - 1.00}}\right) \cdot 1.30$$

$$dc_o = 1.11$$

The curve joining all three dc_o points represents the isodensity line for d = 1.20.

Determination of pore pressure on site using mud logging or wireline data brings the geologist face to face with the problem of choosing one of the above methods : equivalent depth, ratio or Eaton. In fact he has no criteria concerning their accuracy.

Recent statistical studies based on data from 120 wells in three different basins (Gulf of Guinea, Angola, North Sea) have enabled us to determine the areas of application and the accuracy of these methods. These studies involved comparing the calculated values according to each of these methods with actual pressures measured using techniques such as RFT and DST.

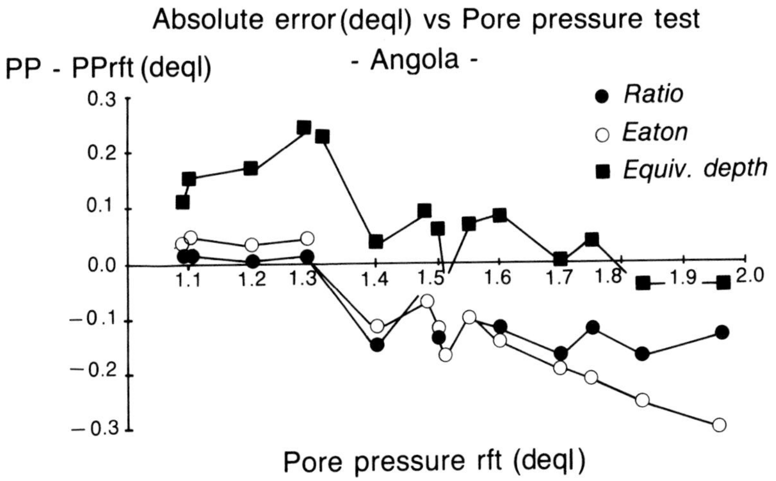

Fig. 128. — Comparative study of different methods for evaluating formation pressure in Angola.

In general, when formation pressures are low ($d_{eql} < 1.40$) the Eaton and ratio methods give the best results (statistical deviations less than 5 density points). On the other hand the equivalent depth method is the most suitable for calculating high pressures ($d_{eql} > 1.40$). Figure 128 illustrates these results in the case of Angola.

It should be noted that the level of accuracy of the equivalent depths method depends directly on the value of the normal compaction trend. The error is greater the smaller the gradient of log "dc" against depth.

4.1.5. *SPECIFIC METHODS*

This section deals with two specific methods of evaluation : the Sigmalog and Prentice's normalised drilling rate. As we shall see, both of these methods are more complicated to use than the above.

4.1.5.1. Sigmalog evaluation

The principle of the Sigmalog has already been described (section 3.4.1.3.).

The evaluation is based on measurement of the differences between the reference trend line $\sqrt{\sigma_r}$ and the average $\sqrt{\sigma_\phi}$ of the $\sqrt{\sigma_o}$ points most to the right. Since $\sqrt{\sigma_\phi}$ is already corrected for the ΔP under normal pressure conditions, it is the value of $\sqrt{\sigma_r}$ (raw sigma corrected for depth only) which is compared with $\sqrt{\sigma_r}$ in order to obtain the ΔP (Fig.129).

The procedure used in calculation is as follows :

— calculate $\sqrt{\sigma_r}$:

$$\sqrt{\sigma_r} = \alpha \cdot \frac{Z}{1\,000} + \beta$$

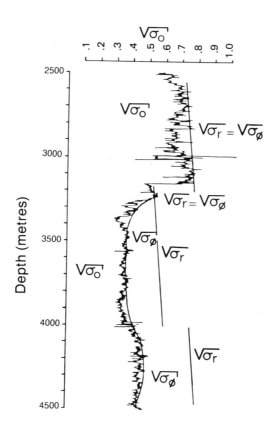

Fig. 129. — Sigmalog trace — representation of $\sqrt{\sigma_o}$ and $\sqrt{\sigma_r}$ (Belloti & Giacca, 1978).

where α = slope of the compaction trend
β = intersection of the trend with $Z = 0$
Z = depth (metres)

— calculate F :

$$F = \frac{\sqrt{\sigma_r}}{\sqrt{\sigma_t}} \text{ (for calculation of } \sqrt{\sigma_t} \text{ see section 3.4.1.3)}$$

— calculate ΔP :

$$\Delta P = \frac{2\,(1 - F)}{1 - (1 - F)^2} \cdot \frac{1}{n}$$

where n = time factor for cleaning the bit face (section 3.4.1.3).

— calculate the equilibrium density corresponding to the pore pressure :

$$d_{eql} = d_m - \frac{\Delta P \times 10}{Z}$$

where d_m = mud density.

These calculations are complex to perform manually, and need to be processed by computer.

The results obtained by the authors of the method appear to be satisfactory for well defined regions (the Po Valley, the Nile Delta). Although its application can be extended to other sectors the lack of any correction for changes in overburden gradient restricts its reliability.

Here again the relationship between the differential pressure and the ratio $\sqrt{\sigma_r}/\sqrt{\sigma_t}$ may be validated geographically or corrected by statistical analyses based on pressure measurements.

4.1.5.2. Normalised drilling rate evaluation (Prentice)

We have already seen a qualitative interpretation of the normalised drilling rate in section 3.4.1.4.

Figure 130 provides an example of its quantitative use, as will be explained below.

A few comments are in order before describing the procedure for calculating pressures on the basis of normalised drilling rate.

— linear scales are used for the plot, the NDR scale being appropriately chosen for the drilling rates expected.
— changes of bit or mud density must be shown by a horizontal line. The normal trend is only considered valid for intervals in which the bit and the mud density are constant. In these circumstances the trend represents compaction and bit dulling. Any deviation reflects changes in lithology or differential pressure.
— as non-clay points are ignored, deviations in NDR from the trend are influenced by ΔP only. (Prentice is of the opinion that the effect of compaction is negligible).

Two empirical relationships between drilling rate and differential pressure have been developed by Prentice. Figure 131a reproduces Prentice's data for South Texas, Figure 131b is based on the data from VIDRINE & BENIT (1968) in South Louisiana (see section 3.4.1.4.).

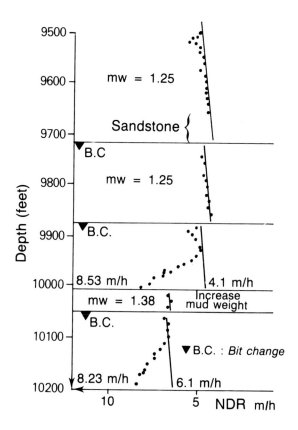

Fig. 130. — Example of a normalised drilling rate (NDR) trace (modified from Prentice, 1980).

There are major differences between these two curves, bearing in mind the similarity of the series concerned. By way of example an increase of 20 % in the drilling rate is equivalent to a reduction of 80 psi in ΔP on curve (a) and a reduction of 300 psi on curve (b) !

It is therefore essential that regional curves for such interpretations should be available in advance if this method is to be used.

Quantitative evaluation includes the following stages :

— calculation of a reference rate NDRR for $\Delta P = 0$
— comparison of NDRR with the observed NDR : NDRR — NDR = ΔNDR
— comparison of ΔNDR with the NDR/ΔP curve for a direct reading of the differential pressure.

Example : see Figure 130 — Use of South Louisiana graph (Fig. 131b).

198

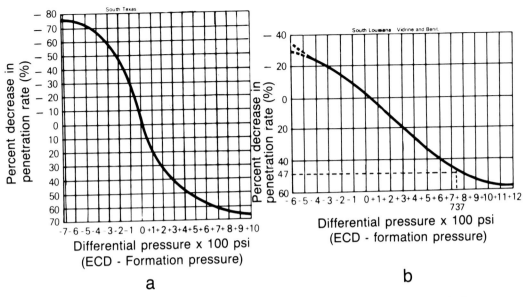

Fig. 131. — Relationship between drilling rate and differential pressure : a) South Texas, b) South Louisiana (from Prentice, 1980).

The formation pressure is normal down to 9 930' (3 027 m) and increases beyond this depth.

$$P_n = 1.08 \, kg \cdot cm^{-2} \cdot m^{-1} \cdot 10$$
$$d \; mud = 1.25 \, kg \cdot cm^{-2}$$

— Calculation of the formation pressure gradient at 10 000' (3 048 m) :

For a normal formation pressure the differential pressure at 3 048 m would be :

$$\Delta P = (d_{mud} - P_n) \times \frac{Z}{10}$$

$$\Delta P = (1.25 - 1.08) \frac{3\,048}{10}$$

$$\Delta P = 51.81 \, kg \cdot cm^{-2} \; (737 \, psi)$$

A ΔP of + 737 psi is equivalent to a reduction in drilling rate of 47 % (Fig.131b).

At 10 000', with theoretical drilling rate deduced from the trend line (d = 1.08) of 4.1 m/h (normal NDR), the reference rate (NDRR) is :

$$NDRR = \frac{normal \; NDR \times 100}{100 - \% \; reduction}$$

199

$$NDRR = \frac{4.1 \times 100}{100 - 47}$$

$$NDRR = 7.7 \text{ m/h}$$

The true NDR is 8.53 m/h, so that the percentage reduction is :

$$\frac{(NDRR-NDR)\ 100}{(NDRR)} = \frac{(7.7 - 8.53)\ 100}{7.7} = -10.78\,\%$$

As this increase in NDR is equivalent to 150 psi (10.5 kg · cm^{-2}) of negative ΔP (Fig. 131b), drilling is proceeding underbalanced.

The formation pressure gradient is calculated as follows :

$$G_P = \frac{\left(d_{mud} \times \dfrac{Z}{10}\right) + \Delta P}{Z/10}$$

$$G_P = \frac{(1.25 \times 304.8) + 10.5}{304.8}$$

$$G_p = 1.284 \text{ kg} \cdot \text{cm}^{-2} \cdot \text{m}^{-1} \cdot 10$$

After increasing the mud density to 1.38 a further displacement in the drilling rate occurs between 10 100 and 10 200′.

The new reference trend is now established for a formation pressure gradient of 1.284.
— Calculation of the formation pressure gradient at 10 200′ (3 109 m) :

$$\Delta P = (1.38 - 1.284) \times \frac{3\ 109}{10} = 29.85 \text{ kg} \cdot \text{cm}^{-2} \text{ (424 psi)}$$

From Figure 131b : 424 psi is equivalent to a 29 % reduction in the drilling rate.
The reference rate at $\Delta P = 0$, given a normal NDR of 6.1 m/h, is :

$$NDRR = \frac{6.1 \times 100}{100 - 29} = 8.59 \text{ m/h}$$

Actual NDR at 10 200 = 8.23 m/h, so that :

$$\% \text{ Reduction} = \frac{8.59 - 8.23}{8.59} \times 100 = 4.1\,\%$$

As a reduction of 4.1 % is equivalent to a positive ΔP of 70 psi (4.9 kg · cm^{-2}) drilling is in equilibrium.

$$\text{Formation pressure gradient} = \frac{\left(1.38 \times \dfrac{3\ 109}{10}\right) - 4.9}{\dfrac{3\ 109}{10}} = 1.364 \text{ kg} \cdot \text{cm}^{-2} \cdot \text{m}^{-1} \cdot 10$$

The curves obtained by Prentice were intended for use only in the regions in which they were established. They can only be applied to other sectors after they have been calibrated and correction coefficients have been established. Ideally they should be redrawn using sufficient statistical data for every region.

Difficulties with correctly placing the bit dulling trend, and the trend shift resulting from bit changes, undeniably place further limitations on the quantitative use of normalised drilling rate.

4.1.6. *EVALUATION BY DIRECT OBSERVATION OF THE DIFFERENTIAL PRESSURE*

The methods of evaluation described above are based on empirical calculations and therefore incorporate major possibilities of error. Direct observation of factors associated with well equilibrium may provide more accurate and reliable information, and is generally the only means of detecting overpressure not directly associated with undercompaction.

● *Gas* : The usefulness of gas shows in qualitative evaluation is described in detail in section 3.4.2.1. We have seen that by monitoring the gas, especially in shales, it is often possible to evaluate well equilibrium satisfactorily. As long as mud weight is close to the equilibrium density, it is possible to monitor background gas, connection gas and the effect of mud weight adjustments on gas shows, so as to achieve satisfactory and continuous evaluation of formation pressure.

● *Mud losses* : Lost circulation may arise for the following two main reasons :

— excessive filtration of mud into a very permeable formation subjected to high differential pressure,

— fracturing of weak horizons (or opening of pre-existing fractures) caused by excessive ΔP (see section 4.3).

Losses may occur while drilling, or be caused by excessive pressure loss due to surging while tripping.

Observing the losses which occur while circulation is in progress, with the well stable under static conditions, provides an accurate picture of well equilibrium. Well balance depends as much on the differential pressure as on the fracture pressure.

It is only safe to use formation pressure data inferred from a mud loss if the location of the zone concerned is accurately known. The loss rate depends not only on the differential pressure but above all on the porosity and permeability of the loss zone, or the nature of the fracture system.

● *Kicks* : A kick indicates that formation pressure is greater than the pressure of the mud column at a given moment. *Only bottomhole kicks should be taken into account for formation pressure evaluation.* Kicks due to gas expansion close to surface are not a direct indication of bottomhole formation pressure.

The kick flow depends on ΔP, the porosity and the permeability of the formation.

With minor kicks, the weight of the mud used for controlling the kick should be close to the equilibrium density. Estimation can be more accurate if the mud weight is adjusted gradually.

If a major kick makes it necessary to shut down the well, formation pressure can be deduced from the shut-in drill pipe pressure :

$$P = \frac{d_{mud} \cdot Z}{10} + P_T$$

where P = formation pressure $(kg \cdot cm^{-2})$
    d_{mud} = mud density $(kg \cdot l^{-1})$
    Z = depth (metres)
    P_T = shut-in drill pipe pressure $(kg \cdot cm^{-2})$

4.1.7. *FORMATION TESTS*

Formation tests can be used for measuring abnormal formation pressure and adjusting mud weight accordingly. They also provide a technique for introducing correction factors into indirect quantitative evaluations. Any large differences between measured and evaluated pressures, might provide information on the origins of the abnormal pressure. If effects such as tectonic stress or thermal expansion exist, pressures may be induced in excess of those due to the overburden effect alone. In addition to this, direct pressure readings can be used to detect abnormal pressure which it was not possible to detect by conventional means while drilling. Note that kicks or losses are natural reflections of disequilibrium, and are "formation tests" in their own right.

Formation tests (DST or production tests) provide only rare, isolated data, and that after the event. They are not designed for the sole purpose of obtaining formation pressure. In fact they are rarely used in abnormally pressured zones (unless there is a hydrocarbon reservoir). In this sense they are generally of little direct use in the detection of abnormal pressure. The information they provide is put to later use, for instance in the preparation of accurate general hydrodynamic maps. These are useful both for information on pressure distribution and for the planning of drilling sites.

On the other hand wireline formation tests (RFT, SFT) are of immediate operational use. Their ability to make a large number of pressure measurements on a single run means they are ideal for monitoring pressure changes. They can be used to resolve doubts about apparent early warnings arising from the use of detection methods while drilling. Because of the way they operate, these measuring techniques require that permeable zones be identified initially through the combined use of drilling rate, lithological description or wireline logs.

These techniques are accurate and reliable. They are, however, limited in practice by the borehole diameter : 6″ to 14 3/4″. This means that when preparing the drilling programme it is necessary to keep in mind that the drilling diameter in the transition zone must be between these limits.

Wireline tests, though costly, are a fundamental part of pressure analysis and, especially in poorly explored zones, may contribute to major reductions in overall drilling costs.

4.2 EVALUATION OF THE OVERBURDEN GRADIENT

Knowledge of the overburden gradient is of prime importance when evaluating formation pressure and fracture gradients.

Because of sediment compaction and the consequent increase in density with depth, the overburden gradient increases rapidly below the surface to reach values which tend to stabilise below a certain depth (Fig. 132).

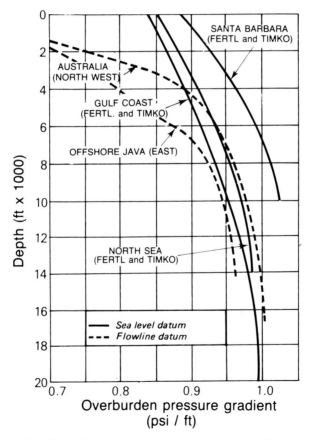

Fig. 132. — Specimen overburden pressure gradients.

Although no significant error arises from assuming a constant gradient in onshore situations, offshore the effects of the depth of water and a greater thickness of poorly consolidated sediments require a more rigorous approach.

Calculation of the overburden gradient, which implies a knowledge of densities, is performed on the basis of wireline logs, "shale" densities (cuttings) or seismic data (interval velocities).

As a reminder, the general formula for calculating overburden pressure is :

$$S = \rho_b \frac{Z}{10}$$

where S = overburden pressure (kg $\cdot$ cm^{-2})
      ρ_b = mean bulk density of the sediment (g $\cdot$ cm^{-3})
      Z = depth (m)

As the mean bulk density is not known for the entire thickness of sediments it is necessary to proceed by means of a cumulated calculation of n intervals I_i(m) of measured bulk densities ρ_i (g $\cdot$ cm^{-3}).

INTERVAL	THICKNESS I_i	APPARENT DENSITY ρ_i	INTERVAL OVERBURDEN PRESSURE S_i kg/cm²	TOTAL OVERBURDEN PRESSURE S kg / cm²	OVERBURDEN GRADIENT GG kg/cm²/10 m	OVERBURDEN GRADIENT kg / cm² / 10 m
0 — 150 (water depth)	150	1.06	15.9	15.9	1.06	
150 — 400	250	1.70	42.5	58.4	1.46	
400 — 700	300	1.80	54.0	112.4	1.61	
700 — 1070	370	1.89	69.9	182.3	1.70	
1070 — 1210	140	2.05	28.7	211	1.74	
1210 — 1400	190	2.02	38.4	249.4	1.78	
1400 — 1780	380	2.20	83.6	333	1.87	
1780 — 1900	120	2.10	25.2	358.2	1.89	
1900 — 2310	410	2.24	91.8	450	1.95	
2310 — 2450	140	2.28	31.9	481.9	1.97	
2450 — 2700	250	2.25	56.3	538.2	1.99	
2700 — 3040	340	2.29	77.9	616.1	2.03	
3040 — 3150	110	2.20	24.2	640.3	2.03	
3150 — 3510	360	2.30	82.8	723.1	2.06	
3510 — 3800	290	2.29	66.4	789 5	2.08	

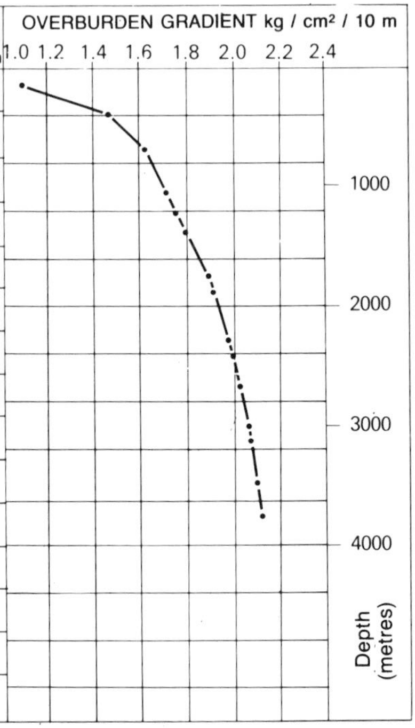

Graph axis labels: 1.0 1.2 1.4 1.6 1.8 2.0 2.2 2.4 ; Depth (metres) markings: 1000, 2000, 3000, 4000

Fig. 133. — Specimen calculation of an overburden gradient.

The formula is then :

$$S = \sum_{i}^{n} \frac{Ii \times \rho_i}{10}$$

where S = overburden pressure (kg · cm^{-2})

Figure 133 is an example of a cumulative calculation of the overburden gradient. The procedure is as follows :

— interval 0-150 m (the water depth), ρ_b = 1.06

Overburden pressure for the interval :

$$S_i = \rho_i \times \frac{Ii}{10} = 1.06 \times \frac{150}{10} = 15.9 \text{ kg} \cdot \text{cm}^{-2}$$

Cumulative overburden pressure :

$$S = Si_n = Si_{n+1}$$
$$S = 0 + 15.9 = 15.9 \text{ kg} \cdot \text{cm}^{-2}$$

Overburden gradient :

$$GG = \frac{\text{cumulative overburden pressure} \times 10}{\text{total depth}}$$

$$GG = \frac{15.9 \times 10}{150}$$

$$GG = 1.06 \text{ kg} \cdot \text{cm}^{-2} \cdot \text{m}^{-1} \cdot 10$$

— Interval 150 — 400 m ; ρ_b = 1.70

Overburden pressure of the interval :

$$S_i = 1.70 \times \frac{250}{10} = 42.5 \text{ kg} \cdot (\text{cm}^{-2})$$

Cumulative overburden pressure : S = 15.9 + 42.5
$$S = 58.4 \text{ kg} \cdot \text{cm}^{-2}$$

Overburden gradient :

$$GG = \frac{58.4 \times 10}{400} = 1.46 \text{ kg} \cdot \text{cm}^{-2} \cdot \text{m}^{-1} \cdot 10 \text{ and so on for the various intervals 400}$$

Plotting the values of the overburden gradient on a graph in relation to depth makes them easier to use. It will be noticed that in this example the gradient is affected by the water depth.

Data selection :

Satisfactory evaluation of the overburden gradient depends on the quality of the data used :

— *Log densities (FDC, LDT, etc.) are the most reliable as long as recording conditions are satisfactory.*

In general some wireline logs (in particular density logs) are not recorded in the upper parts of wells because the hole diameter is too large, the walls are not properly consolidated, and often because there is no hydrocarbon interest in these formations.

The density of the surface sediments can be estimated without too significant an error in the gradient. In some cases an approximation may be obtained from the density of the cuttings.

In order to limit the number of intervals which must be examined the density curve is smoothed in order to eliminate statistical variations. Intervals of uniform density are then identified and measured. It should be noted that isolated low density horizons (<10 m) can be ignored when calculations are done manually. If data processing equipment is available all the variations can be integrated. The results of the two calculation procedures are very similar.

— *Evaluation of the overburden gradient from the sonic log (BELLOTI et al., 1979).*

The fact that density records are only rarely available for an entire borehole has led certain authors to attempt to use sonic log transit times to determine densities.

This method also has the advantage that it can be used in advance of drilling by making use of seismic data, converting interval velocities to transit times.

- Consolidated sandstones

$$\Delta t_{log} = \Delta t_m (1 - \phi) + \Delta t_f \qquad (4.10)$$

where Δt_{log} = transit time reading from the sonic log (μsec/foot)
Δt_m = matrix transit time (μsec/foot)
Δt_f = fluid transit time (μsec/foot)
ϕ = porosity (between 0 and 1)

Formula (4.10) can also be written :

$$\phi = \frac{\Delta t_{log} - \Delta t_m}{\Delta t_f - \Delta t_m} \qquad (4.11)$$

where Δt_f is estimated at 200 μsec/foot.

Values of Δt_m are shown in the table below (Fig. 134) :

Matrix	Δt_m (μsec/foot)
Dolomite	43.5
Limestone	43.5 - 47.6
Sandstone*	47.6 - 55.6
Anhydrite	50
Salt...............................	67
Clay	47 (estimated)

* Δt_m for sandstone varies as a function of the mineralogy of the matrix grains : quartz, feldspar, etc

Fig. 134. — Values of Δt_m.

The following formula, which has been verified by laboratory tests, expresses the relationship between porosity and transit time for consolidated formations :

$$\phi = \frac{\Delta t_{log} - \Delta t_m}{153} \qquad (4.12)$$

- Unconsolidated formations
- — sands :

$$\phi = 1.228 \frac{\Delta t_{log} - \Delta t_m}{\Delta t_{log} + 200} \qquad (4.13)$$

- — clays :

$$\phi = 1.568 \frac{\Delta t_{log} - \Delta t_m}{\Delta t_{log} + 200} \qquad (4.14)$$

The density/porosity relationship is expressed by :

$$\rho_b = \rho_m (1 - \phi) + \rho_f \phi \qquad (4.15)$$

where ρ_b = bulk density reading from the density logs (g · cm^{-3})
ρ_m = matrix density (g · cm^{-3})
ρ_f = fluid density (g · cm^{-3})

The $\rho_b/\Delta t$ relationships are obtained by combining relationships 4.12 or 4.13 and 4.15 :

$$\rho_b = 3.28 - \frac{\Delta t}{89} \text{ for consolidated formations} \qquad (4.16)$$

$$\rho_b = 2.75 - 2.11 \frac{\Delta t - \Delta t_m}{\Delta t + 200} \text{ for unconsolidated formations} \qquad (4.17)$$

Although this method is largely empirical, the authors confirm that use of equation (4.17) yields satisfactory results. Figure 135 illustrates this.

The method for choosing the (Δt) values is different from that for the density log. For each uniform interval selected (in the case of the metric log) :
- — count the number of milliseconds ITT (ticks on the edge of the depth column),
- — multiply the number of milliseconds obtained by 1 000 to obtain a value in µsec,
- — divide by the thickness of the formation (in metres) → µsec/m,
- — multiply by 0.3048 to obtain a value in µsec/foot.

Evaluation of the Δt values from the sonic curve is less accurate but faster. It may be necessary if ITT ticks have not been recorded.

If no seismic data are available for a well, and there are no log data from an adjacent well, densities must be estimated on site by measuring cuttings density.

The techniques for measuring shale densities are described in section 3.4.3.2. and are applicable to all lithologies for this purpose. As the accuracy of cuttings measurement is

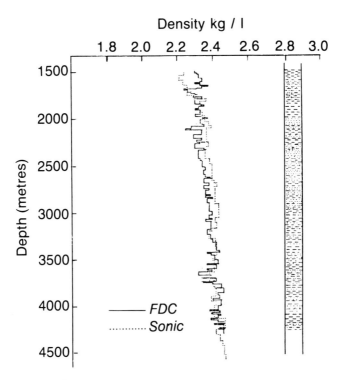

Fig. 135. — Example of a comparison between FDC and sonic calculated densities (Belloti *et al.*, 1979).

variable (due to variations in the proportions of montmorillonite, hydration of the walls, cake, release of confining stresses, etc.) evaluation of pore pressures may be subject to error.

4.3. EVALUATION OF THE FRACTURE GRADIENT

4.3.1. *APPLICATION OF FRACTURE GRADIENT EVALUATION*

In order to prevent kicks while drilling it is necessary to maintain a mud weight such that the mud pressure is slightly higher than the formation fluid pressure at a given level. When penetrating an abnormally pressured zone mud density increases are necessary to maintain equilibrium for the newly drilled formation. This however has several consequences, one of which is that it increases mud pressure throughout the open hole, including in front of previously-drilled fragile zones :

— either because they are porous or already fractured and have a pore pressure which is too low in comparison with the required mud pressure deeper in the hole (these zones may have already been recognised and their pressure will itself place a limit on mud weight, but things are not always so clear-cut). In these circumstances there will be filtration or mud losses.

— or because *the excess mud pressure is sufficient to overcome in situ stresses locally and the geomechanical resistance of the formation,* thus creating fractures. The result is the same : filtration and losses.

It is this latter eventuality which one must attempt to avoid by restricting mud weight to a value below the fracture pressure at the level in question. As we have seen, this is called the "fracture gradient" (section 1.2.2.2).

In fact there is a need to distinguish between the latter gradient and the "leak off gradient" which drillers obtain via the Leak-Off Test (LOT) which is usually carried out just below a previously-run casing shoe prior to drilling ahead, since this is likely to be the weakest point in the next drilling phase. This can roughly be described as injecting mud into the formation, either via the porosity or micro-fractures (see section 4.3.5). Thus, in the presence of any porosity or a weak cement bond of the casing, the LOT will not correspond to the fracture gradient.

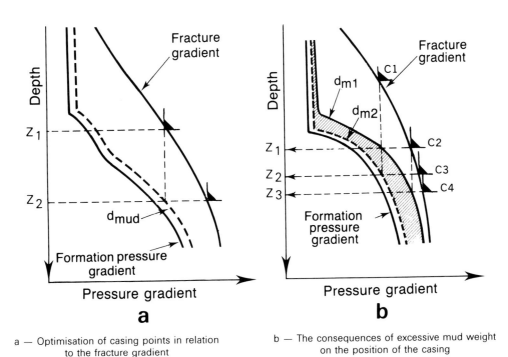

a — Optimisation of casing points in relation to the fracture gradient

b — The consequences of excessive mud weight on the position of the casing

Fig. 136. — Examples of the use of the fracture gradient. d_m = mud density

Information on the fracture gradient is essential (Fig.136) :

— to establish the drilling programme and casing depths. The scheduled mud densities in any one stage should not exceed the lowest expected fracture gradient in the open hole,

— to determine the maximum annular pressure which can be tolerated when controlling kicks, in order to avoid internal blowouts,

— to estimate the pressures required for possible stimulation by hydraulic fracturing.

In this example (Fig.136b) the use of mud weight d_{m1} after setting casing C_1 will require setting further casings C_2 and C_4 whereas a mud weight of d_{m2} only requires casing C_3 to be set.

4.3.2. *BOREHOLE WALL FRACTURE MECHANISM*

In situ formations are subjected to an initial stress condition (see section 1.3). This stress condition is altered in the immediate vicinity of the well. At the wall it depends on, among other things :

— the stress conditions existing in the formation before drilling (in situ stresses),

— the geometry of the hole and its orientation, in particular with respect to the principal in situ stresses,

— mud characteristics (density, rheology, composition, temperature and flow rate),

— the properties of the formation, since a plastic regime may be established at the walls.

Formation fracturing occurs when the stress at the wall exceeds the tensile strength of the rock. The pressure which is in this case exerted on the walls by the mud is called the fracture initiating pressure (FP_1). If the pressure is suddenly reduced, this fracture closes up again. To reopen it, pressure needs only to be increased to value (FP_2). This is lower than the fracture initiating pressure because the fracture already exists. In fact only the perpendicular stresses at the wall hold it closed. FP_2 is the fracture reopening pressure. The fracture can then develop beyond the zone of influence of the well. It will be orientated perpendicular to the minimum component of the in situ stresses. If the hydraulic fluid flow to the well is shut off the pressure will fall and the pressure at which the fracture closes again (FP_3) is taken as an estimate of this minimum in situ stress.

To sum up, the following three "fracture pressures" are distinguished :

FP_1 : fracture initiation pressure (or breakdown pressure)

FP_2 : fracture reopening pressure,

FP_3 : closing pressure.

It is the first two values which are of interest from the point of view of fracture initiation, which is an accident we wish to avoid. These values depend on the hole geometry and drilling conditions. In particular, in a zone having an in situ stress condition such that the horizontal stress is less than the vertical stress the fracture initiating pressure decreases with the inclination of the hole (due to redistribution of the stresses around the hole).

In such an in-situ stress field a highly deviated or even horizontal hole will potentially be subject to greater problems due to fracturing losses than a vertical hole in the same formation. This disadvantage makes it necessary to adopt as the fracture gradient the closing pressure FP_3, which is smaller than the other two (but is above all independent of drilling conditions because it is equal to the minimum component of the in situ stresses).

One should also bear in mind that the value of a fracture pressure measured in a hole is not necessarily characteristic of the formation, but may depend on the drilling conditions, in particular when interpreting leak off tests (see below). Choosing the closing pressure as the fracture gradient also provides a safety margin which, however variable it may be, could still be sufficient. Applying a mud density corresponding to this fracture gradient does not necessarily mean that the formation will fracture.

Finally we should note the importance of the distinction between FP_1 and FP_2. A well which initially resists a density greater than FP_2 may, as a result of a short excess pressure (e.g. surge) above FP_1, no longer be able to resist the same density.

4.3.3. *PRINCIPLES OF FRACTURE GRADIENT EVALUATION*

Evaluating fracture gradient involves evaluating S_3 — the minimum component of the in situ stresses. We saw in chapter 1 that rock deformation and fracture are controlled by the effective stress σ, theoretically defined as the difference between the total stress S and the pore pressure P, and that an estimate of the mean stress supported by the solid matrix can be taken as :

$$\sigma = S - P$$

In this case the minimum effective stress can therefore be defined as

$$\sigma_3 = S_3 - P$$

We have also seen that this stress is usually unknown and could have any value. In most cases the minimum stress is considered to be horizontal, and the equation is conventionally written as :

$$S_3 = \sigma_3 + P = K_3 \sigma + P$$

with σ : effective vertical stress equal to the weight of the overlying sediments
K_3 : ratio of the effective stresses (horizontal to vertical).

Most of the formulae given in the literature arise from consideration of the value of this coefficient K_3. There are two schools :

— the first consists of evaluating K_3 from regional statistical studies of well fracture measurements. This is a good approach in principle (MATTHEWS & KELLY, 1967 ; PILKINGTON, 1978 ; BRECKLES & VAN EEKELEN, 1981) despite its inaccuracies. We shall discuss this below,

— the second makes the assumption that this coefficient has a value which depends on Poisson's ratio μ * for the material in situ, which is regarded as isotropic :

$$K_3 = \frac{\mu}{1 - \mu}$$

This assumption amounts to considering that the formation has not been subjected to any lateral deformation since sedimentation, and that it always deformed elastically during compaction. Not only are these assumptions extremely restrictive, but measurements have often shown that this value of K_3 does not fit. It is now wholly rejected, because it ignores all stages of tectonic history during which the stresses might have changed.

Any method using the isotropic Poisson's ratio for direct estimation of in situ stresses must be rejected, see in particular HUBBERT & WILLIS *(1957),* EATON *(1969),* BIOT *(1955) (who provides an estimate of fracture reopening pressure FP$_2$). Other authors have noted the inaccuracy of this formula and have attempted to modify it.* ANDERSON *et al. (1973) state a relationship between Poisson's ratio and a shale index.* DAINES *(1982) added a tectonic component to the stress determined by Poisson's ratio, thus introducing a further unknown which is without foundation.*

The only valid approach is to study K$_3$ as the unknown in the problem, linking it where appropriate to lithological and regional data.

4.3.4. METHODS OF FRACTURE GRADIENT EVALUATION

The following pages devote space to the main methods for estimating fracture gradient which have caught our attention, independently of the critical study in the previous section. They have been brought together for historical purposes, and to assist readers in dealing with such methods, which are in general use by mud logging companies. The scatter of the results, as will be seen from the case studies (Fig. 144) emphasises the lack of reliability. To close the chapter we have included a theoretical analysis of the fracture gradient (section 4.3.6.). This will be essential reading for anyone wishing to take a less empirical route to the difficult topic of the evaluation of fracture pressures.

4.3.4.1 HUBBERT & WILLIS (1957)

The theoretical basis of formation fracturing given by HUBBERT & WILLIS is that the total stress is equal to the sum of the formation pressure and the effective stress. The authors argue as follows from a theoretical and experimental examination of the mechanics of the hydraulic fracturing of rocks : in situ stresses are characterised by three unequal principal stresses, and hydraulic pump pressures must be approximately equal to the least of these main compressive stresses.

The authors therefore suggest that in geological regions where there are no tangential forces, the greatest stress must be approximately vertical and equal to the overburden

---

* Poisson's ratio : an elastic constant defined as the ratio of the lateral unit strain to the longitudinal strain in a body that has been stressed longitudinally within its elastic limit.

pressure, while the weaker stress must be horizontal and most likely lies between 1/2 and 1/3 of the effective overburden pressure.

The overburden pressure S (total stress) is equal to the sum of the formation pressure (or pore pressure) P and the vertical stress σ_v effectively supported by the matrix.

$$S = P + \sigma_v \tag{4.18}$$

or :

$$\sigma_v = S - P \tag{4.19}$$

Their observations are based on the results of laboratory triaxial compression tests. It is suggested that the pore pressure has no significant effect on the mechanical properties of the rock.

The fracture pressure is defined by the formula :

$$F = \frac{S - P}{3} + P \qquad \text{or} \qquad F = \frac{(S + 2P)}{3} \tag{4.20}$$

$$\text{or} \qquad F = \frac{1}{3}(S - P) + P$$

where F, S and P are measured in $kg \cdot cm^{-2}$. Simply replacing the various pressures with their respective gradient values in $kg \cdot cm^{-2} \cdot m^{-1} \cdot 10\ m$ allows the calculation of the fracture gradient. To keep the text concise all the formulae pertaining to fracture pressures or gradients will use the same terms F, S and P. The reader will employ gradient or pressure values as he wishes.

HUBBERT & WILLIS are sometimes credited with the formula :

$$F = \frac{\mu}{1 - \mu}(S - P) + P \tag{4.21}$$

This version is not given in their article, but taking a Poisson's ratio of 0.25 (see Fig. 142) this formula reduces in effect to equation (4.20).

The authors subsequently considered a Poisson's ratio of 0.5 and amended their formula as a consequence :

$$F = \frac{1}{4} \text{ to } \frac{1}{2}(S - P) + P$$

Measurements of fracture gradients show that the results given by the Hubbert & Willis formula are on the low side.

As this method applies to quite specific regions other authors have attempted to find more widely applicable formulae.

4.3.4.2. MATTHEWS & KELLY (1967)

The MATTHEWS & KELLY method introduces a variable effective stress coefficient into the following formula :

$$F = K_i \, \sigma + P \qquad (4.22)$$

where $K_i = \dfrac{\sigma_h}{\sigma_v}$ effective stress coefficient

This method is original in that measurements of fracturing are used to establish regional empirical curves for K_i (Figure 137a).

Values of K_i were established on the basis of fracture threshold values derived empirically in the field. The effective stress coefficient K_i is variable and depends on depth.

Coefficient K_i is higher for the sands of Southern Texas, which are more clayey than the offshore sands of Louisiana.

The application of this method in zones other than the Gulf Coast requires local variations in K_i to be determined in relation to depth. Because of the relative rarity of these data few regions in the world lend themselves satisfactorily to this type of study.

The fracture gradient is calculated as follows :

— determine the formation pressure gradient (P) and the overburden gradient (S)

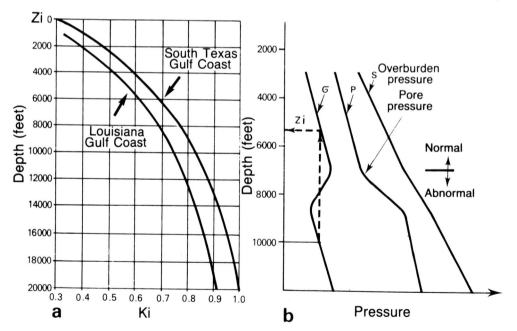

Fig. 137. — Curves for K_i determined by Matthews & Kelly (1967) in South Texas and Louisiana (a) (Courtesy of Oil & Gas Journal) — determination of Z_i : abnormal pressure situations (b)

- calculate the effective stress $\sigma = S - P$
- within an undercompacted zone determine the depth Z_i at which σ is normal (equivalent depth) (Figure 137b)
- read off the value of K_i for Z_i (Figure 137a)
- calculate F using formula (4.22)

4.3.3.3. EATON (1969)

Eaton, stating that rock deformation is elastic, replaces K_i in the above method by a value calculated from Poisson's ratio :

$$F = \left(\frac{\mu}{1-\mu}\right) \sigma + P$$

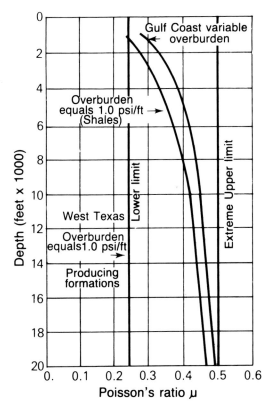

Fig. 138. — Values of Poisson's ratio in relation to the overburden gradient and depth (from Eaton, 1969, courtesy of SPE of AIME).

On the basis that Poisson's ratio and the overburden gradient vary with depth, EATON determined values for Poisson's ratio on the basis of actual regional data for the fracture gradient, the formation pressure gradient and the overburden gradient (Fig. 138).

The use of this method requires regional curves of Poisson's ratio to be established, and it is therefore subject to the same restrictions as Matthews & Kelly's method.

4.3.4.4. ANDERSON *et al.* (1973)

Finding that fracture gradients could vary from one place to another at identical depths in similar formations, ANDERSON *et al.* attributed these variations to the shale content of the reservoirs.

A relationship was established between shale content and Poisson's ratio on the basis of Biot's formula (1955) :

$$\mu = \frac{S - \alpha P}{F + 2S - 3\alpha P} \qquad (4.23)$$

where $\alpha = 1 - Cr/Cb$ Cr = compressibility of solid matrix
$\qquad\qquad\qquad\quad Cb$ = compressibility of porous rock skeleton.
$\quad \alpha$ can be roughly approximated to ϕ_D according to the authors.

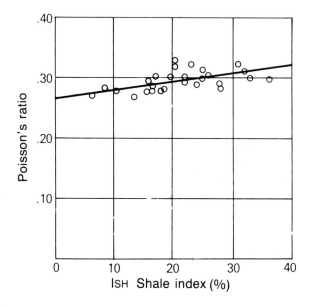

Fig. 139. — The relationship between Poisson's ratio and the shale index
(from Anderson *et al.*, 1973, courtesy of SPE of AIME).

216

On the basis of log data, shale content indexes were established for the measure points in question :

$$Ish = \frac{\phi_S - \phi_D}{\phi_S} \qquad (4.24)$$

where I_{sh} = shale index
ϕ_S = sonic porosity
ϕ_D = density log porosity.

The results obtained yield a linear relationship between Poisson's ratio and the shale index (Figure 139).

The value of μ is obtained using the following equation. A and B are constants defining the straight-line relationship (A = slope, B = y axis intercept) Figure 139.

$$\mu = A\,I_{sh} + B \qquad (4.25)$$

If data are available for overburden gradient, formation pressure gradient, sonic and density logs, then the fracture gradient can be calculated from Biot's formula (4.23) or, as a simplification, from Eaton's formula (4.21).

For this method to be used it is necessary to have sufficient data in advance so as to determine a regional relationship between μ and I_{sh} and to check the validity of the approximation between α and ϕ_D.

Although ANDERSON et al., like EATON, take changes in Poisson's ratio into account, only predominantly sandy lithologies are considered. It should also be remembered (see 4.3.3.) that there are drawbacks to using Poisson's ratio in this way.

4.3.4.5. PILKINGTON (1978)

Returning to previous work in South Texas and Louisiana, PILKINGTON defined an effective stress coefficient K_a which represents the statistical mean of the values of K_i and $\mu/1 - \mu$ used by various authors (Fig. 140).

Substitution of K_i by K_a in Matthews & Kelly's formula (4.22) can be used to calculate the fracture gradient.

Pilkington is of the opinion that these formulae can be applied to Tertiary basins similar to the Gulf Coast, for both normal and abnormal formation pressures. Their use does not extend to brittle rocks such as carbonates, nor to already fractured rocks.

4.3.4.6 CESARONI et al. (1981)

CESARONI et al. emphasised the considerable influence of the mechanical behaviour of rocks on the fracture gradient.

Three situations are highlighted :

— formations with elastic behaviour (sands, sandstones, etc.) with little or no invasion by filtrate due either to low permeability or rapid mud cake formation. In this case the differential

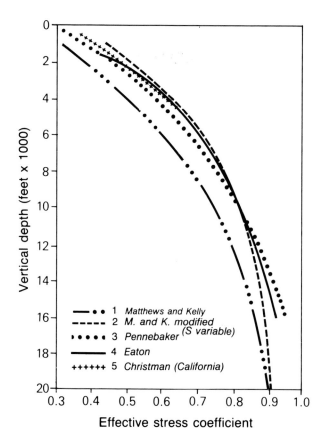

Fig. 140. — Variation with depth of effective stress coefficients given by various authors
(after Pilkington, 1978).
$K_a = 3.9\,S - 2.88$ for $S \leqslant 0.94$ psi/foot
$K_a = 3.2\,S - 2.22$ for $S > 0.94$ psi/foot

pressure is entirely supported by the borehole wall.

$$F = \frac{2\,\mu}{1 - \mu}\,\sigma + P \tag{4.26}$$

— elastic formations with deep invasion (eg. coarse, poorly cemented sands).

$$F = 2\mu\sigma + P \tag{4.27}$$

— plastic formations (shale, marl, salt, etc.)

$$F = S \tag{4.28}$$

218

μ is evaluated at 0.25 for clean sands and unfractured carbonates at shallow depth and at 0.28 for shaly sands, carbonates and sandstones at greater depth. Generally formula 4.26 is used except where very thick shales, salt or coarse sand are encountered, in which case formulae 4.27 and 4.28 may be used.

4.3.4.7. BRECKELS & VAN EEKELEN (1981)

Using fracture gradient data not only from the Gulf Coast but also from Venezuela, Brunei and the North Sea, BRECKELS & VAN EEKELEN established relationships between the minimum horizontal stress S_3 and depth.

The results are very comparable between basins : Figure 141.

The relationship between S_3/depth/pore pressure (for the Gulf Coast) :
$S_3 = 0.053 \, Z^{1.145} + 0.46 \, (P - P_n)$ for $Z \leqslant 3\,500$ m
$S_3 = 0.264 \, Z - 317 + 0.46 \, (P - P_n)$ for $Z > 3\,500$ m

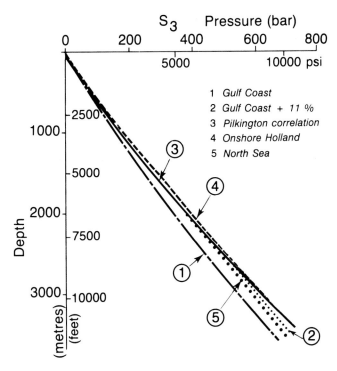

Fig. 141. — Relationship between S_3 and depth for different basins (from Breckels & Van Eekelen, 1981, courtesy of SPE of AIME).

where S_3 = minimum horizontal stress (bar)
  Z = depth (metres)
  P = pore pressure (bar)
  P_n = normal pore pressure (bar).

4.3.4.8. DAINES (1982)

DAINES, taking up the work of Eaton, introduced a superimposed tectonic stress correction.

$$F = \sigma_t + \sigma \left(\frac{\mu}{1 - \mu}\right) + P \qquad (4.29)$$

where σ_t = superimposed tectonic stress.

The value of σ_t may be evaluated from the first leak off test on drilling. It is regarded as being constant for the rest of the well.

In addition to the effect of tectonic stresses, DAINES emphasises the role of lithology in the calculation of fracture gradient. The Poisson's ratios below (Fig. 142) were obtained from experiments on the propogation of sound (shear waves).

Clay	0.17 to 0.50	— medium	0.06
Conglomerate	0.20	— fine	0.03
Dolomite	0.21	— poorly sorted, shaly	0.24
Limestone		— fossiliferous	0.01
— micritic	0.28	Shale	
— sparitic	0.31	— calcareous	0.14
— porous	0.20	— dolomitic	0.28
— fossiliferous	0.09	— siliceous	0.12
— argillaceous	0.17	— silty	0.17
Sandstone		— sandy	0.12
— coarse	0.05 to 0.10	Siltstone	0.08

Fig. 142. — Poisson's ratios for different lithologies (Daines, 1982).

Figure 143 shows an example of the application of Daines' method for the calculation of the fracture gradient F, incorporating corrections for tectonic stress and the changes in Poisson's ratio due to lithology.

The values of Poisson's ratio are only indicative and cannot be checked on site in view of the difficulties in identifying the arrival of shear waves. New multiple transmitter/receiver sonic logging tools such as EVA (ELF/CGG) and SDT (Schlumberger) may allow in situ determination of Poisson's ratio.

In addition to this the determination of σ_t from a leak off test is approximate, especially as the value of μ chosen for the calculation may be erroneous.

220

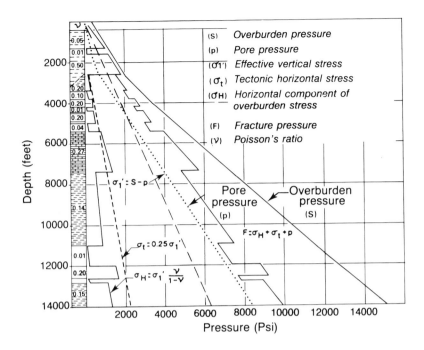

Fig. 143. — Depth/fracture pressure chart in relation to lithology (at constant pore pressure gradient) (Courtesy of EXLOG).

Finally we should remember that the division of the stress field into an elastic component and a tectonic stress is illusory.

4.3.4.9. BRYANT (1983)

The Bryant method is limited to modifying K_i in the Matthews & Kelly formula to take account of the contribution of pore pressure to the mechanical properties of the matrix.

— if $P < 1.4\ P_n$, Matthews & Kelly's data for South Texas are used (Figure 137)
— if $P > 1.4\ P_n$, K_i is obtained from $K_i = \dfrac{P}{S}$.

• Comparison of results from previous methods

The methods for evaluating fracture gradient have been applied to wells with abnormal pressure (Fig. 144). When compared with measurements obtained by LOT, the results appear highly scattered.

221

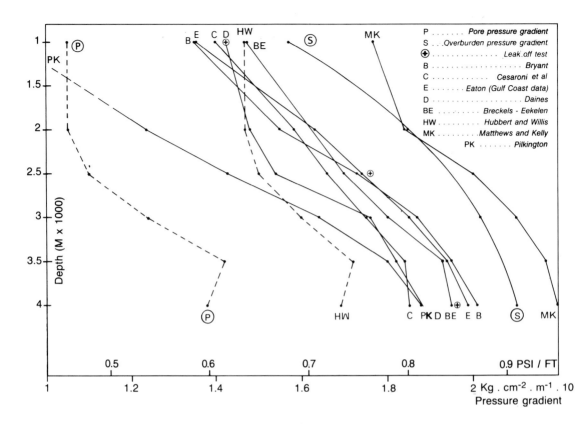

Fig. 144. — Comparison between the different methods of calculating fracture gradient.

These methods must therefore be used with considerable caution. A theoretical analysis of fracture gradient which takes account of all relevant parameters is proposed below (section 4.3.6).

4.3.5. *LEAK-OFF TEST*

A leak-off test (LOT) involves increasing the mud pressure in a shut-in well until mud is injected into the formation.

It is carried out to determine the maximum pressure which can be applied to the formation while drilling the next phase after running casing. LOT's are generally carried out after cementing casing, redrilling cement and shoe, and drilling a few metres of formation. This zone is assumed to be the most fragile part of the future open hole. The data obtained

222

may be supplemented by a further LOT when a permeable zone has been penetrated for the first time. In fact we have already seen that the most fragile section is the part in which the stress condition and the pressure gradient are weakest. The maximum permissible mud pressure therefore depends equally on the latest value from the LOT and the change in both these parameters.

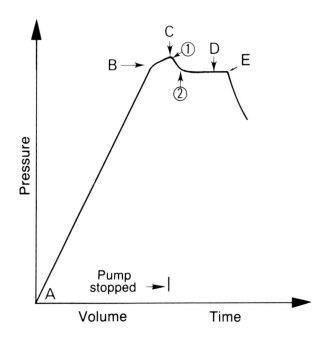

Fig. 145. — Example of a typical record from a leak off test.

The figure above is a typical example of a LOT recording. It is interpreted as follows :

A-B : linear increase in annular pressure proportional to volume pumped, corresponding to the elastic behaviour of the formation.

B : shoulder of curve corresponding to start of leak off = *LOT pressure.*

B-C : reduced increase in pressure per volume pumped, mud penetrating the formation.

C : pumping stopped. Two situations may arise : either a pressure plateau or *leak-off pressure* is reached (1) or there is a sudden drop in pressure (2) following wall breakdown or reopening of a previously created or natural vertical fracture in the wall.

C-D : fracture propogation ceases, pressure falls to a stabilised pressure D which is less than or equal to pressure B.

E : pressure purge — end of test.

223

When the excess pressure is purged the amount of mud recovered should be equal to the volume pumped. If the pressure at D is lower than at B it is likely that the cracks will remain partially open, obstructed by cuttings or mud. In this case the amount of mud recovered will be less than that pumped. In permeable zones this may give rise to major losses arising from enlargement of the area of contact between mud and the formation.

LOT therefore runs the risk of weakening the walls of the hole thus reducing the fracture gradient. In order to avoid this problem, predetermined maximum values can be assumed to be sufficient in the light of the expected pressures, so that formation breakdown pressure is not reached (integrity test). In this case these values are default values and as such cannot be used to evaluate fracture gradients.

4.3.6. *THEORETICAL ANALYSIS OF THE FRACTURE GRADIENT*

Through failing to take into account the importance of mud-induced tangential stresses in the fracture gradient estimation process, many authors have been led into assuming that the fracture pressure is equal to the total horizontal stress, whereas in fact this is only rarely true. Thus it takes only the inclusion of thermal stress phenomena to explain the improvements in LOT values often observed during drilling. This was often attributed to a process enigmatically called "clay healing".

The theoretical argument which appears below is the outcome of recent research, and has been specially written for this book by Maury & Guenot of Elf Aquitaine's Rock Mechanics Department.

4.3.6.1. Fracture Orientation

The orientation and position of a fracture in a well depends on the stress field at the wellsite and the orientation of the hole relative to that field. Taking the case of a vertical well, two situations are possible depending upon the local stresses. If the minor component is vertical, the fracture in the borehole wall will be perpendicular to it, and therefore horizontal. Otherwise, with the minor component horizontal, the wall fracture will be vertical, in the direction of the stronger of the two horizontal components. We will deal with the latter case, which is felt to be the more common.

4.3.6.2. Stress state in the borehole walls

When a vertical fracture opens in the borehole wall the stress perpendicular to this vertical face, which we will call the tangential stress, is cancelled out. This wall stress is the result of :

— the concentration of in-situ stresses around the hole,
— the pressure applied to the wall by the mud,
— stresses of thermal origin,

— stresses linked with mud filtration conditions in the formation.

The last three drilling-related factors must therefore be estimated in order to deduce the in situ stresses, which are the real unknowns in the problem.

Let us assume to start with that the formation is porous, but that filtration has stopped due to the presence of mud cake. It has been demonstrated that under elastic conditions, if S_2 and S_3 are the horizontal components of the in-situ stresses, the tangential stress at the wall (S_θ) will be a minimum in the direction of S_2 and will be equal to :

$$S_\theta = 3\ S_3 - S_2$$

and using the notation previously employed (section 4.3.3.) :

$$S_3 = K_3\ \sigma_v + P_F$$
$$S_2 = K_2\ \sigma_v + P_F$$

$$S_\theta = (3\ K_3 - K_2)\ \sigma_v + 2\ P_F$$

where P_F is the formation pore pressure and σ_v is the vertical component of the effective stresses in situ.

Let us assume, as a simplification, that the horizontal components are equal ($K_3 = K_2 = K$). We obtain :

$$S_\theta = 2\ (K\ \sigma_v + P) = 2\ K\ (\rho_b - \rho_w)\ Z + 2 Z \rho_w$$

where ρ_b is the mean density of the overlying formations
ρ_w is the pore pressure gradient and Z is the depth in question.

Mud of density ρ_m has the effect of reducing the tangential stress in the wall at level Z.

$$\Delta S_{\theta m} = -\ \rho_m\ Z$$

As it circulates the mud cools or heats the wall of the well depending upon its position in the open hole. This alters the tangential stress much more than is usually thought :

$$\Delta S_{\theta T} = \frac{\alpha\ E\Delta T}{1 - \mu} = Z F_T$$

where α : expansion coefficient of the rock
ΔT : thermal disturbance in the wall
E,ν : elastic characteristics of the rock in situ
F_T : thermal factor.

The use of E (Young's modulus) and ν is justified here because it is assumed that drilling subjects the rock to a deformation which remains elastic.

The order of magnitude of this stress, for a rock of average rigidity, is :

$\alpha = 0.8 \cdot 10^{-5}\ (°C)^{-1}$
$E = 500\ 000\ kg \cdot cm^{-2} \rightarrow \Delta S_{\theta T} = 5.7\ \Delta T$
$\mu = 0.3$

For example, for a cooling of $\Delta T = 40°C$ the thermal stresses are :

$$\Delta S_{\theta T} = - 230 \text{ kg} \cdot \text{cm}^{-2}$$

which is equivalent, at around 2 000 m, to a change of some $1.2 \text{ g} \cdot \text{cm}^{-3}$ in the mud density. We will call the gradient which results from these thermal stresses the thermal factor, F_T. In general F_T is negative, but it may happen that it is slightly positive in top hole.

If a regime of permanent filtration from the well to the formation is set up this also changes the tangential stress :

$$S_{\theta h} = \frac{1 - 2\mu}{1 - \mu} (\rho_m - \rho_w) Z$$

It therefore remains to determine the pore pressure in the wall of the well to obtain the effective tangential stress at the wall, and to compare this with the tensile strength of the rock σ_T. If there is no filtration the pore pressure $\rho_w Z$ is taken. Otherwise the mud pressure $\rho_m Z$ is taken to be the pressure at the wall. This yields an overall effective tangential stress :

A) *No filtration (clays, unfractured carbonates)* :

$$\left(\frac{\sigma\theta}{Z}\right) = 2K (\rho_b - \rho_w) + 2\rho_w - \rho_m + F_T - \rho_w = 2K\rho_b + \rho_w (1 - 2K) - \rho_m + F_T$$

B) *Filtration (sandstone, sands, etc.)*

$$\left(\frac{\sigma\theta}{Z}\right) = 2K (\rho_b - \rho_w) + 2\rho_w - \rho_m + F_T + \frac{1 - 2\mu}{1 - \mu} (\rho_m - \rho_w) - \rho_m$$

or $\quad$ $$\left(\frac{\sigma\theta}{Z}\right) = 2K (\rho_b - \rho_w) + \frac{\rho_w - \rho_m}{1 - \mu} + F_T$$

The following two formulae provide an expression for the fracture initiation gradient of a vertical fissure in the wall.

$$F_A = 2K \rho_b + \rho_w (1 - 2K) + F_T + \sigma T/Z$$

or,

$$F_B = \left[2K (\rho_b - \rho_w) + F_T + \frac{\sigma T}{Z}\right] (1 - \mu) + \rho_w$$

If a measurement of the fracture initiation gradient, F, is available, it can be interpreted on the basis of both assumptions in order to obtain a lateral stress coefficient K_A or K_B :

$$\left(\frac{\sigma\theta}{Z}\right)_A = 2K_A \rho_b + \rho_w (1 - 2K_A) - F + F_T = - \frac{\sigma T}{Z}$$

or

$$2K_A (\rho_b - \rho_w) = F - \rho_w - F_T - \frac{\sigma T}{Z}$$

and

$$K_A = \frac{F - F_T - \rho_w - \sigma_T/Z}{2 (\rho_b - \rho_w)}$$

Similarly with filtration assumption B we obtain :

$$K_B = \frac{F - \rho_w}{2 (\rho_b - \rho_w) (1 - \mu)} - \frac{F_T + (\sigma_T/Z)}{2 (\rho_b - \rho_w)}$$

It is clear that determination of the lateral stress by this method is subject to uncertainties which may be resolved with experience :

— the presence or absence of filtration,
— the value of the tensile strength of the borehole wall,
— the value of the thermal disturbance.

4.3.6.3. Numerical example

Let us consider a clay horizon and a sandstone horizon at 2 000 m depth. The mechanical characteristics are as follows :

	Clay	Sandstone
Young's Modulus E	100 000 kg · cm^{-2}	600 000 kg · cm^{-2}
Poisson's Ratio ν	0.20	0.25
Tectonic stress Δ_T	50 kg · cm^{-2}	70 kg · cm^{-2}
Expansion coefficient α	8.10-6(°C)$^{-1}$	1.2 . 10-5(°C)$^{-1}$
Thermal stress $\dfrac{\alpha E}{(1 - \mu)}$	1.00 kg · cm^{-2} · °C^{-1}	9.6 kg · cm^{-2} · °C^{-1}

The mean density of the overlying sediments evaluated by density log or cuttings measurement is :

$$\rho_b = 2.50$$

The formation pressure gradient estimated by means of the methods described in the previous chapter is :

$$\rho_w = 1.15$$

Initially we will assume that the in-situ stress condition is known, in order to see the effect of the various hypotheses on the variation in the fracture gradient :

$$K_2 = 0.9$$
$$K_3 = 0.7$$

or :

$$K = \frac{3K_3 - K_2}{2} = 0.6$$

The fracture gradient FP_3, as defined previously, is thus equal to the minor horizontal component of the in situ stresses, or : $FP_3 = K3 (S - P) + P$

$$giving \ FP_3 = 0.7 (2.50 - 1.15) + 1.15 = 2.10$$

If it is assumed that the in situ stress condition is identical in both formations this value will be unique to the formations under scrutiny.

As far as the fracture initiating pressure FP_1 or breakdown pressure is concerned, we will assume that filtration occurs in the sandstone and that no filtration occurs in the clay.

We will also study the situation where fracturing occurs a little while after a circulation in the well and where the formation has not regained thermal equilibrium, so that :

$\Delta T = -15°C$ for clay (less conducting)
$\Delta T = -10°C$ for sandstone.

Applying the above formulae, bearing in mind the assumptions mentioned, yields the following values for fracture initiation pressure FP_1 and reopening pressure FP_2 :

	With tensile strength FP_1		Zero tensile strength FP_2		
	Without ΔT	With ΔT	Without ΔT	With ΔT	FP_3
Sandstone	2.62	2.26	2.36	2.00	2.10
Clay	3.02	2.94	2.77	2.70	2.10

The very strong effect of thermal stresses, even for low cooling values, can be seen. As a result of this thermal load it is possible to obtain a higher initiation pressure in a LOT made at the shoe but at the end of a drilling phase than in a test made right at the start of the phase because in the latter case cooling of the section in question, close to the bottom of the hole, is greater.

Initiation pressure is subject to major uncertainties unless the test is performed after circulation has been stopped for a prolonged period and with reduced flow.

Taking the problem the other way round, ie. from measured values of the fracture pressures, values of K can be obtained from the previous formulae :

$K_A = 0.38 - 0.37 \ F_T$ for clays
$K_B = 0.507 - 0.37 \ F_T$ for sands

and if the same cooling is taken as before we obtain :

$K_A = 0.41$
$K_B = 0.68$

On the other hand if we had (incorrectly) used the relationship whereby the closure pressure can be obtained directly from the horizontal stress we would have obtained :

$K = 0.78$

4.3.7. *CONCLUSION*

— *The fracture gradient cannot be estimated correctly except by appropriate regional interpretation of hydraulic pressure limit measurements performed on the initial wells in a field, in terms of the in situ stresses.*

— *When carrying out a LOT it is necessary to evaluate whether the fracture gradient obtained relates to leak-off into the formation, a leak-off limit (pressure plateau), a formation breakdown (pressure drop), a propagation pressure or a closing pressure.*

— *Thermal and hydraulic stresses must be taken into account in the interpretation if one wishes to obtain a true value of the minimum in-situ stress. Only this can be used to explain the anomalies in LOT commonly found when drilling (apparent improvement in LOT pressure with circulation time).*

— *The fracture gradient (initiation or re-opening) depends on the nature of the formation but also on the well geometry. Thus it is possible to drill a vertical well with no problems and encounter mud losses in a deviated well in the same zone with the same mud weight.*

5. — SUMMARY AND OVERALL CONCLUSIONS

Although undoubted progress has been made in the understanding of abnormal pressure in recent years, there has been no fundamental change for several decades in the way drilling problems are perceived and solved. Take for example the "d exponent" introduced in 1966 by JORDEN & SHIRLEY on the basis of work published by BINGHAM in 1964. It remains to this day one of the most reliable undercompaction detection techniques available, and has survived almost unmodified in all that time. Insight into the origins of abnormal pressure has progressed essentially as a result of laboratory experiments conducted under conditions which are sometimes quite different from those in the field. Significant progress will only be made in the future if all the theories, hypotheses and conclusions produced in these laboratories can be tested against geological reality.

However this may be, it stands out from the exhaustive list of the origins of abnormal pressure provided in this handbook, that an imbalance between the rate of subsidence and the capacity for fluid expulsion represents the most widespread cause of pressure anomalies. The outstanding feature of this essentially dynamic process is the temporary nature of overpressures, which exist for finite periods of geological time and which vary with the geological setting. Recent deposits, such as delta regions on passive continental margins and accretion prisms in subduction trenches, are the sites where abnormal pressure is most likely to develop. Other contributory factors to abnormal pressure (clay and sulfate diagenesis, osmosis, tectonics etc.) should on the whole be regarded as secondary, even though one cause or another may have appeared to play an important role for a time, depending on the prevailing state of research. In fact several causes may occur together. Any analysis must take into account phenomena which are seldom fully understood, such as the expulsion of very large volumes of water in the course of compaction, diagenetic phenomena, or the prevailing stress field.

The most important thing of all for the subsurface geologist, however, is the concept of a transition zone. The absence of such a zone is an additional difficulty in the early detection of abnormally pressured zones. This is why it is essential to prepare adequately for every drilling programme. Techniques such as geophysics, sedimentology and tectonics must be used, together with the examination of reference wells, to study the regional geological context and to establish with reasonable certainty whether abnormal pressure exists, and if so define its characteristics.

On the other hand, the presence of a transition zone associated with undercompaction phenomena lends itself to the use of a whole range of detection methods. First of all there are predictive techniques, which are currently capable of determining pressure profiles with suitable accuracy. They lead to improved projection of casing points, and in so doing help to achieve drilling objectives. Then there is a growing range of methods, which can be used in combination, for detection while drilling. It should also be kept in mind that measurements while drilling (M.W.D.) will make a useful contribution to this field in years to come.

Even with all these techniques available to personnel at the drilling site and the increasingly important contribution from data processing, the experience of the wellsite geologist and the driller will still be needed for a long time yet to overcome the many difficult problems associated with drilling through zones of abnormal pressure.

HELP CHART

HELP CHART

FOR THE PREPARATION AND SURVEILLANCE OF WELLS WITH ABNORMAL PRESSURE RISKS

WELL PREPARATION

| Unexplored zone. Risk analysis : |

▷ Determine the type of deposit, the tectonic context
   (sedimentological, geophysical techniques, etc.)

▷ Determine its ability to retain abnormal pressure
   in relation to :

| Undercompaction Basic conditions : — rapid subsidence — reduced drainage capacity |

- type of deposit : delta, platform, etc.
- lithological nature of the formations : clays, evaporite deposits, etc.
- rate of sedimentation / subsidence
- drainage capacity, spatial distribution of drains (current directions and sand body orientation)
- continuity / discontinuity of drains
- age of the formations

▷ Attempt to determine the possible origins of abnormal pressure
   (if the above criteria are borne out) :

| Do not overlook the possibility of laterally induced pressure |

- subsidence rate / drainage capacity ratio
  ⇒ risk of undercompaction
  at the same time consider the lithology and permeability of the overlying cover, the role of sealing and/or capacity for undercompaction
- envisage other possibilities :
  • diagenetic contributions in relation to the nature of the deposits : smectite dehydration, gypsum
  • thermal expansion of water and hydrocarbons

| Explored zone : |

▷ Investigation of reference wells : attempt to understand origin and distribution of
   abnormal pressure

- isobath maps of the top of the undercompaction
- piezometric / potentiometric maps
- tectonic maps
- wireline log studies / correlations
- pressure maps : geographical distributions, vertical changes
- programs for the computer processing of wireline logs

▷ Geophysical studies

- well to well correlations
- seismic facies analyses, nature of reflections, frequencies
- estimation of the sand / clay percentage
- calculation of interval velocities, preparation of a pseudo-sonic profile

▷ Estimation of pore pressures (equivalent depth method applied to pseudo-sonic)
   and fracture pressures : pressure / depth diagrams.

WELL SITE SURVEILLANCE

First order parameters :

▶ Drilling rate / d exponent
(avoid tool wear corrections except in well known sectors where statistical studies
may be carried out)

▷ Determine the normal compaction trend in clays
(bear in mind changes in sand, silt content, etc.)
— observe the factors which might affect the trend :
change of bit, wear, etc.
When approaching an undercompacted zone its identification
will be easier if the drilling parameters or bit type are not
changed appreciably

Do not have blind
confidence in
forecasts : bear all
parameters in mind
while drilling and
avoid preconceived ideas

▷ Observe any negative deflection : if this is not due to an external cause there
is a presumption of abnormal pressure which should be reinforced by a
change in at least one other detection parameter

▶ Gas :

Inform and establish
a good working
relationship with
the tool-pusher

1) in clays/shales :
- increase in background gas
- appearance, increase in connection gas,
- change in the composition of gas shows : relative increase in heavy components
2) after passing though a reservoir :
- gas % does not return to its previous value

▶ Lithology : beyond a certain clay thickness (regionally defined)
▶ Torque while drilling
▶ Drag when pulling string
▶ Changes in the volume of circulating mud : pit level, flow measurement

Second order parameters :

▶ Shale density : care must be taken with cuttings selection and data quality
▶ Mud Temperature : a potentially useful tool, but rarely effective
▶ Shale factor : should be used in sectors where traditional parameters are poorly reliable
▶ Cuttings : volume, size, shape, presence of cavings
▶ Miscellaneous (see text) : local applications

Determination of pressure while drilling :

1) By estimation :

- from the d exponent : equivalent height, ratio, Eaton methods ; requires a good
approximation to the normal compaction trend
- from an intermediate sonic log

Check balance of deql / deqv by the reaction
of gas to changes in mud weight

2) By direct measurement : intermediate RFT runs to check estimates and
achieve best adjustment of mud weight

Readjustment of pressure analyses while drilling :

▶ Intermediate logs
▶ Seismic check : checkshots, VSP

Difficult exploration wells :

▶ Consider the use of M.W.D. (Gamma Ray + Formation resistivity)

6. — APPENDICES

6.1. SI UNITS, CONVERSION FACTORS AND TABLES (*)

The SI International System of Units is built up from seven base units.

TABLE 1 — BASE SI UNITS

PHYSICAL QUANTITY	NAME	SYMBOL
Length..	metre	m
Mass..	kilogram	kg
Time	second	s
Electric current	ampere	A
Thermodynamic temperature	kelvin	K
Amount of substance...	mole	mol
Luminous intensity	candela	cd

Units derived from the base units are shown as algebraic expressions using the mathematical signs for multiplication (.) and division (/).

TABLE 2 — EXAMPLES OF SI UNITS DERIVED FROM THE BASE UNITS

PHYSICAL QUANTITY	UNIT		EXPRESSED IN BASE OR SUPPLEMENTARY UNITS (*)
	NAME	SYMBOL	
SPACE AND TIME			
area, surface	square metre	m^2	m^2
volume	cubic metre	m^3	m^3
angular velocity..........................	radian per second	rad/s	$s^{-1}.rad$
velocity	metre per second	m/s	$m.s^{-1}$
acceleration	metre per second squared	$m.s^{-2}$	$m.s^{-2}$
frequency	hertz	Hz	s^{-1}
frequency of rotation	second to the power minus one	s^{-1}	s^{-1}
MECHANICAL			
density	kilogram per cubic metre	kg/m^3	$m^3.kg$
weight flow	kilogram per second	kg/s	$kg.s^{-1}$
volume flow rate........................	cubic metre per second	m^3/s	m^3s^{-1}
momentum	kilogram-metre per second	kg.m/s	$m.kg.s^{-1}$
kinetic moment	kilogram-square metre/second	$kg.m^2/s$	$m^2.kg.s^{-1}$
moment of inertia......................	kilogram-square metre	$kg.m^2$	$m^2.kg$
force	newton	N	$m.kg.s^{-2}$
moment of force........................	newton-metre	N.m	$m^2.kg.s^{-2}$
pressure, stress.........................	pascal	Pa	$m^{-1}.kg.s^{-1}$
viscosity (absolute)....................	pascal-second	Pa.s	$m^{-1}.kg.s^{-2}$
kinematic viscosity	square metre per second	m^2/s	$m^2.s^{-1}$
surface tension	newton per metre	N/m	$kg.s^{-2}$
energy, work, quantity of heat	joule	J	$m^2.kg.s^{-2}$
power, energy flux	watt	W	$m^2.kg.s^{-3}$

(*) All units can be expressed in base units or supplementary units.

(*) *Source :* Guide pratique pour le Système International d'Unités (SI) (Practical guide to the SI International System of Units) — MOUREAU, M. — Technip ed., 1980.

CONVERSION FORMULAE

LENGTH

1 inch		=	2.540×10^{-2}m	2.540 cm
1 foot	= 12 inches =		3.048×10^{-1}m	30.48 cm
1 yard (yd)	= 3 feet =		0.9144 m	
1 fathom	= 6 feet =		1.8288 m	
1 mile (statutory)		=	1609 m	1.609 km
1 mile (nautical)		=	1852 m	1.852 m

SURFACE AREA

1 square inch	=	645×10^{-6}m^2	645 mm^2
1 square foot	=	0.092 m^2	929 cm^2
1 square yard	=	0.836 m^2	
1 acre = 9 square feet	=	4047 m^2	0.4047 ha
1 square mile	=	2.589×10^6 m^2	2.589 km^2

VOLUME

equals 1	US gallon	Imperial gallon (UK)	litre	barrel (Am.)	cubic metre	cubic foot
US Gallon	1	0.8327	3.7853	0.0238	0.0038	0.1337
Imperial Gallon (UK)	1.201	1	4.5459	0.0286	0.0045	0.1605
Litre	0.264	0.220	1	0.0063	0.001	0.0353
Barrel (Am.)	42	34.973	158.984	1	0.159	5.6154
Cubic metre	264.17	219.97	1 000	6.2898	1	35.3147
Cubic foot	7.481	6.2288	28.317	0.1781	0.0283	1

MASS

1 ounce (Oz)	=	2.83×10^{-2} kg	28.35 grams
1 pound (lb)	=	0.453 kg	453.59 grams

PRESSURE

1 pound force per square foot (psf) = 4.788 Pa

1 pound force per square inch (psi) = 6.895×10^3 Pa

TEMPERATURE

− 17.8 ºC	=	0º Fahrenheit
0 ºC	=	32º F
100 ºC	=	212º F
Formulae ºC	=	$\dfrac{5}{9}$(ºF − 32)
ºF	=	$\dfrac{9}{5}$(ºC + 17,8)

CONVERSION FACTORS TO SI OR ASSIMILATED UNITS

TO CONVERT FROM	TO	MULTIPLY BY
acre	m²	4.047×10^3
acre	ha	0.4047
acre-foot	m³	1.233×10^3
angström	m	1.0×10^{-10}
atmosphère	Pa	1.013×10^5
bar	Pa	1.0×10^5
barrel (42 gallons)	m³	0.1590
barrel per foot	m² (m³/m)	5.216×10^{-1}
barrel per inch	m² (m³/m)	6.259
barrel per long ton	m³/kg	1.565×10^{-4}
barrel per short ton	m³/kg	1.752×10^{-4}
barye	Pa	1.0×10^{-1}
Btu (International Table)	J	1.055×10^3
Btu (lt)	kWh	2.930×10^{-4}
Btu (mean)	J	1.056×10^3
Btu (thermochemical)	J	1.054×10^3
Btu (IT) per barrel	J/m³	6.636×10^3
Btu (IT) per gallon	J/m³	2.321×10^5
Btu (IT) per gallon (US)	J/m³	2.787×10^5
Btu (IT) per hour	W	2.931×10^{-1}
Btu (IT) per minute	W	1.758×10^1
Btu (IT) per second	W	1.055×10^3
bushel	m³	3.523×10^{-2}
calorie (International Table)	J	4.187
calorie (IT)	kWh	1.163×10^{-6}
calorie (mean)	J	4.190
calorie (thermochemical)	J	4.184
calorie (IT) per pound	J/kg	9.230
centimeter of mercury	Pa	1.333×10^3
centimeter of water	Pa	98,06
centipoise	Pa.s	1.0×10^{-3}
cheval-vapeur	W	735.4
degree (angle)	rad	1.745×10^{-2}
denier (international)	kg/m	1.0×10^{-7}
dram	kg	1.772×10^{-3}
dram (troy)	kg	3.888×10^{-3}
dram (US fluid)	m³	3.697×10^{-6}
dyne	N	1.0×10^{-5}
electron volt	J	1.602×10^{-19}
erg	J	1.0×10^{-7}
fathom	m	1.829
fluid ounce (US)	m³	2.957×10^{-5}
foot	m	0.3048
square foot	m²	9.290×10^{-2}
cubic foot	m³	2.832×10^{-2}
cubic foot per barrel	m³/m³	0.178
cubic foot per pound	m³/kg	6.243×10^{-2}

TO CONVERT FROM	TO	MULTIPLY BY
foot pound	J	1.356
foot pound	kWh	3.766×10^{-7}
frigorie (negative kilocalorie)	J	4.187×10^3
furlong	m	2.012×10^2
gallon (imp)	m^3	4.546×10^{-3}
gallon (US dry)	m^3	4.404×10^{-3}
gallon (US liquid)	m^3	3.785×10^{-3}
gallon (US liquid) per minute	m^3/h	0.227
gauss	T	1.0×10^{-4}
gill (US)	m^3	1.183×10^{-4}
grain	kg	6.479×10^{-5}
gram	kg	1.0×10^{-3}
horsepower	W	7.457×10^2
horsepower (boiler)	W	9.809×10^3
horsepower (electric)	W	7.460×10^2
horsepower (hydraulic)	W	7.46×10^2
horsepower (metric)	W	7.355×10^2
horsepower hour	J	2.685×10^6
hundredweight (long)	kg	5.080×10^1
hundredweight (short)	kg	4.536×10^1
inch	m	2.54×10^{-2}
inch mercury (32° F)	Pa	3.386×10^3
inch water (32° F)	Pa	2.490×10^2
inch pound force	N/m	1.130×10^{-1}
kilocalorie	kWh	1.163×10^{-3}
kilogramme force	N	9.806
kilogramme force/m^2	Pa	9.806
kilowatt-hour	J	3.6×10^6
kilogrammeter	J	9.806
kip	N	4.45×10^3
kip per square foot	Pa	4.788×10^4
knot (international)	m/s	0.5144
league (British nautical)	m	5.556×10^3
league (statute)	m	4.878×10^3
light year	m	9.460×10^{15}
link	m	2.012×10^{-1}
litre	m^3	1.0×10^{-3}
mho	s	1
micron	m	1.0×10^{-6}
mil	m	2.54×10^{-6}
mile (US nautical)	m	1.852×10^3
mile (US statute)	m	1.609×10^3
mile (US statute) per hour	m/s	4.470×10^{-3}
square mile	m^2	2.590×10^6
millibar	Pa	100.0
millimeter of mercury	Pa	1.333×10^2
oersted	A/m	79.58
ounce (US fluid)	m^3	2.957×10^{-5}
ounce force	N	0.278
ounce mass	kg	2.835×10^{-2}
ounce mass (troy)	kg	3.110×10^{-2}

TO CONVERT FROM	TO	MULTIPLY BY
peck (US)	m³	8.809×10^{-3}
pennyweight	kg	1.555×10^{-3}
perch (rod, pole)	m	5.029
pint (US dry)	m³	5.506×10^{-4}
pint (US liquid)	m³	4.732×10^{-4}
poise	pa.s	1.0×10^{-1}
pound force	N	4.448
pound force per square foot	Pa	4.788×10^{1}
pound force per square inch (psi)	Pa	6.895×10^{3}
pound mass	kg	0.4536
pound mass (troy)	kg/m³	0.3732
pound mass per barrel	kg/m³	2.853
pound mass per cubic foot	kg	1.602×10^{1}
psi see : pound force per square inch		
psi per foot	Pa/m	2.262×10^{4}
poundal	N	0.138
quart (US dry)	m³	1.101×10^{-3}
quart (US liquid)	m³	9.463×10^{-4}
quarter section (160 acres)	m²	6.475×10^{5}
		(approx, 65 ha)
rod	m	5.029
roentgen	C/kg	2.579×10^{-4}
second (angle)	rad	4.848×10^{-6}
section (640 acres)	m²	2.590×10^{6}
		(approx. 260 ha)
slug	kg	14.59
span	m	0.2286
stokes	m²/s	1.0×10^{-4}
therm	J	1.05×10^{8}
thermie (= 1 megacalorie)	kWh	1.163
thermie	J	4.186×10^{6}
ton (long, 2240 pounds)	t	1.016
ton (short, 2000 pounds)	t	0.9071
torr	Pa	1.333×10^{2}
yard	m	0.9144
square yard	m²	8.361×10^{-1}
cubic yard	m³	7.645×10^{-1}

EQUIVALENCE BETWEEN DENSITY UNITS

Specific gravity	Gradient (psi per thousand feet of depth)	Pounds per gallon
0.78	338	6.5
0.84	364	7.0
0.90	390	7.5
0.96	416	8.0
1.00	433	8.3
1.02	442	8.5
1.08	468	9.0
1.14	494	9.5
1.20	519	10.0
1.26	545	10.5
1.32	571	11.0
1.38	597	11.5
1.44	623	12.0
1.50	649	12.5
1.56	675	13.0
1.62	701	13.5
1.68	727	14.0
1.74	753	14.5
1.80	779	15.0
1.86	805	15.5
1.92	831	16.0
1.98	857	16.5
2.04	883	17.0
2.10	909	17.5
2.16	935	18.0
2.22	961	18.5
2.28	987	19.0
2.34	1013	19.5
2.40	1039	20.0
2.46	1065	20.5
2.52	1091	21.0
2.58	1117	21.5
2.64	1143	22.0
2.70	1169	22.5
2.76	1195	23.0
2.82	1221	23.5
2.88	1247	24.0

EQUIVALENCE BETWEEN A.P.I. GRAVITY AND DENSITY

CONVERSION FORMULAE
(at a temperature of 60ºF, 15.56ºC)

$$\text{Gravity in degrees API} = \frac{141.5}{\text{density}} = 131.5$$

$$\text{Density} = \frac{141.5}{\text{API gravity} + 131.5}$$

API GRAVITY	DENSITY	BARRELS/ METRIC TONNE
0	1,0760	5.86
10	1.0000	6.30
15	0.9659	6.52
18	0.9465	6.66
20	0.9340	6.75
22	0.9218	6.84
24	0.9100	6.93
26	0.8984	7.02
28	0.8871	7.10
30	0.8762	7.19

API GRAVITY	DENSITY	BARRELS/ METRIC TONNE
32	0.8654	7.28
34	0.8550	7.37
36	0.8448	7.46
38	0.8348	7.55
40	0.8251	7.64
42	0.8155	7.73
44	0.8063	7.82
46	0.7972	7.91
48	0.7883	8.00
50	0.7796	8.09

API GRAVITY	DENSITY	BARRELS/ METRIC TONNE
55	0.7587	8.31
60	0.7389	8.53
65	0.7201	8.76
70	0.7022	8.98
75	0.6852	9.20
80	0.6690	9.43
85	0.6536	9.65
90	0.6388	9.87
95	0.6247	10.10
100	0.6112	10.32

SIMPLIFIED AND APPROXIMATE CONVERSIONS
1 metric tonne crude oil — 7.3 barrels
1 barrel crude — 0.14 metric tonne
1 barrel/day — 50 t/year
1 cubic foot — 30 litres or 1/35 m^3
1 cubic foot/day — 10m^3/year

SALINITY - DENSITY RELATIONSHIP

(NaCl content)

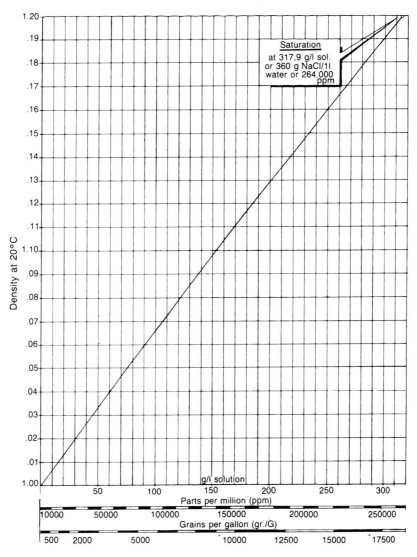

CONVERSION FROM DEGREES CELSIUS
TO DEGREES FAHRENHEIT AND VICE - VERSA

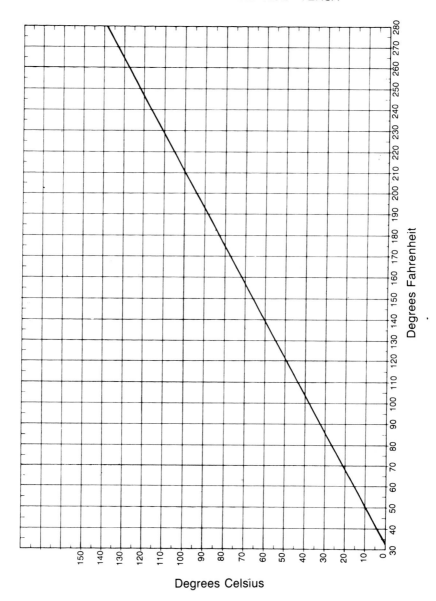

Degrees Fahrenheit

Degrees Celsius

6.2. REFERENCES

ANDERSON, R.A., INGRAM, D.S. & ZANIER, A.M. (1973). — Determining fracture pressure gradients from well logs. — *J. Petroleum Technol.,* **25,** 11, 1259-1268.

ANDERSON, D.M. & LOW, P.F. (1958). — Density of water adsorbed by lithium, sodium and potassium-bentonite — *Soil Sci. Soc. Am. Proc.,* **22,** 97-103.

BARKER, C. (1972). — Aquathermal pressuring. Role of temperature in development of abnormal pressure zones. — *Bull. amer. Assoc. petroleum Geol.,* **56,** 10, 2068-2071.

BELLOTTI, P. & GERARD, R.E. (1976). — Instantaneous log indicates porosity and pore pressure. — *World Oil,* **183,** 9, 90-94.

BELLOTTI, P. & GIACCA, D. (1978). — Sigmalog detects overpressures while drilling deep wells. — *Oil and Gas J.,* **70,** 35, 148-150.

BELLOTTI, P., DI LORENZO, V. & GIACCA, D. (1979). — Overburden gradient from sonic 1og. Trans. SPWLA, Londres, mars 1979.

BINGHAM, M.G. (1964). — A new approach to interpreting rock drillability (3). — *Oil and Gas J.* — **62,** 46, 173-179.

BIOT, M.A. (1955). — Theory of elasticity and consolidation for a porous anisotropic solid. — *J. Apply. Phys.,* **26,** 2, 115-135.

BOURGOYNE, A.T. & YOUNG, F.S. (1973). — A multiple regression approach to optimal drilling and abnormal pressure detection. — *6th SPE of AIME drilling rock mech. conf. preprint, SPE 4238,* 91-106.

BRECKELS, I.M. & VAN EEKELEN, H.A.M. (1981). — Relationship between horizontal stress and depth in sedimentary basins. — *56th SPE of AIME, Fall technical conference, San Antonio, Texas, SPE 10336,* 21 p.

BRUCE, C. (1973). — Pressured shale and Related Sediment Deformation : Mechanism for development of regional contemporaneous faults. — *Bull. amer. Assoc. petroleum Geol.,* **57,** 5, 878-886.

BRYANT, T.M. (1983). — Fracture gradient techniques — A review — Baroid Logging Systems course notes.

BURST, J.F. (1969). — Diagenesis of gulf coast clayey sediments and its possible relation to petroleum migration. — *Bull. amer. Assoc. petroleum Geol.,* **53,** I, 73-93.

CESARONI, R., GIACCA, D., SCHENATO, A. & THIERREE, B. (1981). — Determining Frac Gradients while drilling. — *Petroleum Eng.,* **53,** 7, 60-86.

CESARONI, R., GIACCA, D., POSSAMAI, E. & SCHENATO, A. (1982). — IEOC Experience in overpressure detection and evaluation in the mediterranean offshore. Agip document presented at the EGPC production seminary, Cairo, Nov. 1982.

CHAMBRE SYNDICALE DE LA RECHERCHE & DE LA PRODUCTION DU PETROLE ET DU GAZ NATUREL (1979). — Prévention et maitrise des éruptions — Editions TECHNIP, Paris.

CHAPMAN, R.E. (1974). — Clay diapirism and overthrust faulting. — *Bull. geol. Soc. Amer.* **85**, 10, 1597-1602.

CHAPMAN, R.E. (1980). — Mechanical versus thermal cause of abnormally high pore pressures in shales. — *Bull. amer. Assoc. Petroleum Geol.*, **64**, 12, 2179-2183.

CHIARELLI, A., SERRA, O., GRAS, C., MASSE, P. & TISON, J. (1973). — Etude automatique de la sous-compaction des argiles par diagraphies différées, méthodologie et applications. — *Rev. Inst. Franç. Pétrole*, **28**, 1, 19-36.

CHIARELLI, A. (1973). — Etude des nappes aquifères profondes. Contribution de l'hydrogéologie à la connaissance d'un bassin sédimentaire et à l'exploration pétrolière. — *Thèse Sc. Univ. Bordeaux*, 1-244.

CHIARELLI, A. & DUFFAUD, F. (1980). — Pressure origin and distribution in Jurassic of Viking basin. — *Bull. amer. Assoc. Petroleum Geol.*, **64**, 8, 1245-1266.

CLAYPOOL, G.E., KAKLAN, I.R. & PRESLEY, B.J. (1973). — Gas analyses in sediment samples from legs 10, 11, 13, 14, 15, 18 and 19 (DSDP) — Book, US Gov. printing office, Washington, DC; **19**, 879-884.

COMBS, G.D. (1968). — Prediction of pore pressure from penetration rate. *S.P.E. of AIME, 43rd AIME Fall Meeting, Houston, Texas, SPE 2162*, 16 p.

CORRE, B., EYMARD, R. & GUENOT, A. (1984). — Numerical Computation of Temperature Distribution in a Wellbore While Drilling — *SPE of AIME, 59th Annual Technical Conference, Houston, Texas, SPE 13208*, 12 p.

CRANS, W., MANDL, G. & HAREMBOURE, J. (1980). — On the theory of growth faulting : a geomechanical delta model based on gravity sliding. — *J. Petroleum Geology*, **2**, 3, 265-307.

CUNNINGHAM, R.A. & EENINK, J.G. (1958). — Laboratory study of effect of overburden, formation and mud column, pressures on drilling rate of permeable formations. — *SPE of AIME, 33rd Annual Fall Meeting, Houston, Texas.*, 9 p.

DAINES, S.R. (1982). — Prediction of fracture pressures for wildcat wells. *J. Petroleum Technol.*, **34**, 4, 863-872.

DAVIS, D., SUPPE, J. & DAHLEN, F.A. (1983). — Mechanics of fold-and-thrust belts and accretionary wedges. — *J. Geophys. Res.*, **88**, 82, 1153-1172.

DICKINSON, G. (1953). — Geological aspects of abnormal reservoir pressures in Gulf Coast Louisiana. — *Bull. amer. Assoc. petroleum Geol.*, **37**, 2, 410-432.

DUNOYER de SEGONZAC, G. (1964). — Les argiles du Crétacé supérieur dans le bassin de Douala (Cameroun) : problèmes de diagenèse. — *Bull. Serv., Carte géol. Als. — Lorr.*, **17**, 4, 287-310.

EATON, B.A. (1969). — Fracture gradient prediction and its application in oil field operations. — *J. Petroleum Technol.*, **21**, 1353-1360.

EATON, B.A. (1972). — Graphical method predicts geopressures worldwide. *World Oil*, **182**, 6, 51-56.

ECKEL, J.R. (1958). — Effect of pressure on rock drillability. *SPE of AIME, Trans.*, **213**, 1-6.

EVAMY, B.D., HAREMBOURE, J., KAMERLING, P., KNAAP, W.A., MOLLOY, F.A. & ROWLANDS, P.H. (1978). — Hydrocarbon habitat of Tertiary Niger Delta. — *Bull. amer. Assoc. Petroleum Geol.*, **62**, 1, 1-39.

ESPITALIE, J., MARQUIS, F. & BARSONY, I. (1984). — Geochemical logging by the oil show analyser, in Analytical Pyrolysis. Butterworth Ed. London.

EXPLORATION LOGGING (1981). — The pressure Log Reference Manual.

FERTL, W.H. (1976). — Abnormal formation pressures, developments in Petroleum Science no 2. — Elsevier Ed., Amsterdam, 382 p.

FERTL, W.H. & TIMKO, D.J. (1970). — Occurrence and significance of abnormal pressure formations. — *Oil and Gas J.,* **68**, 1, 97-108.

FERTL, W.H. & TIMKO, D.J. (1971). — Utilisation des diagraphies en recherche, forage et production d'hydrocarbures pour l'étude des problèmes de pressions géostatiques. — *Rev. Inst. franç. Pétrole,* **26**, 9, 687-714.

FERTL, W.H. & TIMKO, D.J. (1972-1973). — How down hole temperatures, pressures affect drilling. — *World Oil,* Jan. 72 to Mars 73, **175**, 4, 45-50.

FRIPIAT, J. & LETELLIER, M. (1984). — Microdynamic behaviour of water in clay gels below the freezing point. — *J. Magnetic Resonance,* **57**, 2, 279-286.

GALLE, E.M. & WOODS, H.B. (1963). — Best constant bit weight and rotary speed for rotary rock bits. *Drilling and production practice (ApI),* 48-73.

GEERTSMA, J. (1961). — Velocity log interpretation : the effect of rock bulk compressibility. — *Soc. Petroleum Eng. J.,* **14**, 235-348.

GOGUEL, J. (1969). — Le rôle de l'eau et de la chaleur dans les phénomènes tectoniques. — *Rev. géog. phys. géol. dyn.,* **11**, 2, 153-164.

GOLDSMITH, R.G. (1975). — It's simple : drilling rates predict mud weights needed. — *World Oil,* **181**, 6, 100-102.

GRAF, D.L. (1982). — Chemical osmosis, reverse chemical osmosis, and the origin of subsurface brines. — *Geochim. Cosmochim. Acta,* **46**, 8, 1431-1448.

GRETENER, P.E. (1969). — Fluid pressure in porous media, its importance in geology : A review. — *Can. Soc. Petroleum Geol.,* **17**, 3, 255-295.

GRETENER, P.E. (1977) (revised 1981). — Pore pressure : Fundamentals, general ramifications and implications for structural geology. — *Bull. amer. Assoc. Petroleum Geol.,* Education Course Note Series ≠ 490 p.

HANSHAW, B.B. & ZEN, E.A. (1965). — Osmotic equilibrium and overthrust faulting. — *Bull. geol. Soc. Amer.* **76**, 12, 1379-1386.

HARKINS, K.L. & BAUGHER, J.W. (1969). — Geological significance of abnormal formation pressures. — *J. Petroleum Technol.,* **21**, 8, 961-966.

HASSAN, M. & HOSSIN, A. (1975). — Contribution à l'étude des comportements du thorium et du potassium dans les roches sédimentaires — *C.R. Acad. Sci. Paris,* **280**, Série D, 5, 533-535.

HEDBERG, H.D. (1974). — Relation of methane generation of undercompacted shales, shale diapirs, and mud volcanœs. — *Bull. amer. Assoc. petroleum Geol.,* **58**, 4, 661-673.

HEIM, A. (1878). — Mechanismus der Gebirgsbildung. — *Rev. polytech. Suisse,* **24**.

HOTTMAN, C.E. & JOHNSON, R.K. (1965). — Estimation of formation pressures from log-derived shale properties. — *J. Petroleum Technol.,* **17**, 6, 717-722.

HUBBERT, M.K. & RUBEY, W.W. (1959). — Role of fluid pressure in mechanics of overthrust faulting, mechanics of fluid filled porous solids and its application to overthrust faulting. — *Bull. geol. Soc. Amer.* **70**, 2, 115-166.

HUBBERT, M.K. & WILLIS, D.G. (1957). — Mechanics of hydraulic fracturing. *Trans. AIME,* **210**, 153-168.

JAUZEIN, A. (1974). — Les données sur le système $CaSO_4$, H_2O et leurs implications géologiques. — *Rev. Géogr. Phys. Géol. Dyn.,* **16**, 2, 151-160.

JONAS, J., BROWN, D. & FRIPIAT, J. (1982). — NMR study of kaolinite water system at high pressure. — J. Colloid Interface Science, 89, 2, 374-377.

JONES, P.H. (1967). — Hydrology of Neogene deposits in the Northern Gulf of Mexico basin. — Proc. Ist Symp. Abnormal Subsurface Pressure, Louisiana, State Univ. pp.91-207.

JORDEN, J.R. & SHIRLEY, O.J. (1966). — Application of drilling performance data to overpressure detection. — *J. Petroleum Technol.,* **28**, 11, 1387-1394.

KARTSEV, A.A. (1971). — The principal stages in the formation of petroleum. 8th World Petroleum Congress, Moscow, **2**, 3-11.

KERN, R. & WEISBROD, A. (1964). — Thermodynamique de base pour minéralogistes, pétrographes et géologues. MASSON ed. Paris, 243 p.

KHARAKA, Y.K. & BERRY, F.A.F. (1973). — Simultaneous flow of water and solutes through geologic membranes. — *Geochim. Cosmochim. Acta,* **37**, 12, 2577-2603.

LEWIS, C.R. & ROSE, S.C. (1970). — A theory relating high temperatures and overpressures. — *J. Petroleum Technol.* **22**, 11-16.

LOUDEN, L.R. (1971). — " Chemical caps" can cause pressure build up. — *Oil and Gas J.,* **69**, 46, 144-146.

LUCASEAU, F. & LE DOUARAN, S. (1985). — The blanketing effect of sediments in basins formed by extension : numerical model. Applications to the Gulf of Lion and Viking graben — *Earth planetary Science Letters* — **74**, 1, 92-102.

LUHESHI, M.N. (1983). — Estimation of formation temperature from borehole measurements — *Geophys. J.R. astron. Soc.,* **74**, 3, 747-776.

MAGARA, K. (1974). — Compaction, ion filtration, and osmosis in shale and their significance in primary migration. — *Bull. amer. Assoc. petroleum Geol.,* **58**, 2, 283-290.

MAGARA, K. (1975a). — Importance of aquathermal pressuring effect in Gulf Coast. — *Bull. amer. Assoc. petroleum Geol.,* **59**, 10, 2037-2045.

MAGARA, K. (1975b). — Reevaluation of montmorillonite dehydration as cause of abnormal pressure and hydrocarbon migration — *Bull. amer. Assoc. Petroleum Geol.,* **59**, 2, 292-302.

MAGARA, K. (1978). — Compaction and fluid migration-developments in petroleum Science n° 9. — Elsevier Ed., 33l p.

MAGARA, K. (1981). — Fluid dynamics for cap-rock formation in Gulf Coast. *Bull. amer. Assoc. Petroleum Geol.,* **65**, 7, 1334-1343.

MARENKO, Y.I. & POSTNIKOV, V.G. (1967). — Causes of abnormally high rock pressure in the Osinskii zone of Markovo oil field — Neftegazov. — *Geol. Geofiz.* **10**, 10-12.

MASCLE, A., LAJAT, D. & NELY, G. (1979). — Sediments of formation linked to subduction and to argilokinesis in the southern Barbados Ridge from multichannel seismic surveys. IVth Latin American Geological Congress, Port of Spain, Trinidad, in press.

MATHEWS, W.R. & KELLY, J. (1967). — How to predict formation pressure and fracture gradient. — *Oil and Gas J.,* **65**, 8, 92-106.

MORELOCK, J. (1967). — Sedimentation and mass physical properties of marine sediments, Western Gulf of Mexico — *Thèse, Texas A and M. University* (University Microfilms, Ann Arbor, Mich., 156 p.)

NWACHUKWU, S.O. (1976). — Approximate geothermal gradients in Niger delta sedimentary basin. — *Bull. amer. Assoc. Petroleum Geol.,* **60**, 7, 1073-1077.

OLSEN, H.W. (1972). — Liquid movement through kaolinite under hydraulic, electric and osmotic gradients. — *Bull. amer. Assoc. petroleum Geol.,* **56**, 10, 2022-2028.

OUTMANS, H.D. (1959). — The effect of some drilling variables on the instantaneous rate of penetration. — *SPE of AIME Reprint series,* n° 6, 102-114.

OVERTON, H.L. & TIMKO, D.J. (1969). — The salinity principle — a tectonic stress indicator in marine sands. — *Log. Anal.,* **10**, 3, 34-43.

PENNEBAKER, E.S. (1968). — Seismic data indicate depth, magnitude of abnormal pressure. — *World Oil,* **166**, 7, 73-78.

PERRODON, A. (1980). — Dynamics of oil and gas accumulations. — *Bull. Centres Rech. Explor.-Prod. Elf-Aquitaine,* Mem. **5**; 368 p.

PERRY, D. (1970). — Early diagenesis of sediments and their interstitial fluids from the continental slope, northern Gulf of Mexico. *Gulf Coast Assoc. Geol. Soc.,* **20**, 219-227.

PERRY, E. & HOWER, J. (1972). — Burial diagenesis in Gulf Coast pelitic sediments. — *Clays Clay Miner.,* **18**, 3, 165-177.

PILKINGTON, P.E. (1978). — Fracture gradient estimates in tertiary basins. *Petroleum Engineer International.,* **50**, 5, 138-148.

POWERS, M.C. (1959). — Adjustment of clays to chemical charge and the concept of the equivalence level. — *Clays Clay Miner.,* **2**, New-York, Pergamon Press, 309-326.

POWERS, M.C. (1967). — Fluid-release mechanisms in compacting marine mudrocks and their importance in oil exploration. — *Bull. amer. Assoc. Petroleum Geol.,* **51**, 7, 1240-1254.

PRENTICE, C.M. (1980). — Normalized penetration rate predicts formation pressure. — *Oil and Gas J.,* **78**, 32, 103-106.

REHM, B. & CLENDON, R.Mc (1971). — Measurement of formation pressure from drilling data. — SPE of AIME, *46th Annual Fall Meeting. New Orleans, Louisiane, SPE 3601,* 12 p.

REYNOLDS, E.B. (1973). — The application of seismic techniques to drilling techniques. — *SPE of AIME, 48th Fall Meeting, Las Vegas, Nevada, SPE 4643,* 8 p.

ROBINSON, L.H., SPEERS, J.M. WATKINS, L.A., BARRY, A. & MILLER, J.F. (1980). — Exxon MWD tool yields unexpected downhole data. — *Oil and Gas J.,* **78**, I6, 86-90.

ROUCHET, J. (du) (1978). — Eléments d'une théorie géomécanique de la migration de l'huile en phase constituée. — *Bull. Centres Rech. Explo. Prod. Elf-Aquitaine,* **2**, 2, 337-373.

RUBEY, W.W. & HUBBERT, M.K. (1959). — Role of fluid pressure in mechanics of overthrust faulting. Overthrust belt in geosynclinal area of western Wyoming in light of fluid pressure hypothesis. — *Bull. amer. Assoc. Petroleum Geol.,* **70**, 2, 167-206.

SCHOLLE, P.A. (1978). — Porosity prediction in shallow versus deep water limestones-primary porosity preservation under burial conditions. *53rd. Annual Technical Conf. of SPE Houston, Texas, Preprint SPE 7554,* 6 p.

SHARP, J.M. (1983). — Permeability controls on aquathermal pressuring. *Bull. amer. Assoc. petroleum Geol.,* **67**, 11, 2057-2061.

SPEED, R.C. (1983). — Structure of the accretionary complex of Barbados. *Bull. geol. Soc. Amer.* **94**, 1, 92-116.

TERZAGHI, K. (1923). — Die Berechnung der Durchlass igkeitsziffer des Tones aws dem Verlanf der Hydrodynamischen Spannungserscheinungen — *Sb. Akad. Wiss. Wien,* 132-135.

TERZAGHI, K. & PECK, R.B. (1948). — Soil mechanics in engineering practice. J. Wiley, New-York. — 566 p.

TISSOT, B.P. & WELTE, D.H. (1978). — petroleum formation and occurrence. a new approach to oil and gas exploration. — Springer, Berlin, 538 p.

VALERY, P., NELY, G., MASCLE, A., BIJU-DUVAL, B., LE QUELLEC, P. & BERTHON, J.L. (1985). — Structure et croissance d'un prisme d'accrétion tectonique proche d'un continent : la ride de la Barbade au sud de l'arc antillais — Géodynamique des Caraïbes, Symposium Paris, 5-8 février 1985, TECHNIP ed., Paris.

VIDRINE, D.J. & BENIT, E.J. (1968). — Field verification of the effect of differential pressure on drilling rate. — *SPE of AIME 42th Annual Fall Meeting, Houston, Texas, SPE 1859*, 11 p.

WESTBROOK, G.K., SMITH, M.J., PEACOCK, J.H. & POULTER, M.J. (1982). Extensive underthrusting of undeformed sediment beneath the accretionary complex of the Lesser Antilles subduction zone — *Nature,* **300**, 5893, 625-628.

WESTBROOK, G.K. & SMITH, M.J. (1983). — Long decollements and mud volcanœsEvidence from the Barbadas Ridge Complex for the role of high pore fluid pressure in the development of an accretionary complex. *Geology,* **11**, 5, 279-283.

WOODSIDE, W. & MESSMER, M.H. (1961). — Thermal conductivity of porous media. — *J. Appl. Physics,* **39**, 1688-I706.

YOUNG, R. & LOW, P.F. (1965). — Osmosis in argillaceous rocks. — *Bull. amer. Assoc. Petroleum Geol.,* **49**, 7, 1004-1008.

ZOELLER, W.A. (1978). — Instantaneous log is based on surface drilling data. — *World Oil,* **187**, 1, 97-105.

ZOELLER, W.A. (1984). — Determine pore pressures detection from M.W.D. gamma ray logs. — *World Oil,* **198**, 4, 97-102.

6.3. SUBJECT INDEX

L O U I S - J E A N
avenue d'Embrun, 05003 GAP cedex
Tél. : 04.92.53.17.00
Dépôt légal : 578 — Août 2001
Imprimé en France